Springer Series in

ADVANCED MICROELECTRONICS 23

Springer Series in

ADVANCED MICROELECTRONICS

Series Editors: K. Itoh T. Lee T. Sakurai W.M.C. Sansen D. Schmitt-Landsiedel

The Springer Series in Advanced Microelectronics provides systematic information on all the topics relevant for the design, processing, and manufacturing of microelectronic devices. The books, each prepared by leading researchers or engineers in their fields, cover the basic and advanced aspects of topics such as wafer processing, materials, device design, device technologies, circuit design, VLSI implementation, and subsystem technology. The series forms a bridge between physics and engineering and the volumes will appeal to practicing engineers as well as research scientists.

18 **Microcontrollers in Practice**
By I. Susnea and M. Mitescu

19 **Gettering Defects in Semiconductors**
By V.A. Perevoschikov and V.D. Skoupov

20 **Low Power VCO Design in CMOS**
By M. Tiebout

21 **Continuous-Time Sigma-Delta A/D Conversion**
Fundamentals, Performance Limits and Robust Implementations
By M. Ortmanns and F. Gerfers

22 **Detection and Signal Processing**
Technical Realization
By W.J. Witteman

23 **Highly Sensitive Optical Receivers**
By K. Schneider and H.K. Zimmermann

24 **Bonding in Microsystem Technology**
By J.A. Dziuban

Volumes 1–17 are listed at the end of the book.

K. Schneider H. Zimmermann

Highly Sensitive Optical Receivers

With 191 Figures and 25 Tables

Springer

Dipl.-Ing. Dr. techn. Kerstin Schneider
Univ. Professor Dr.-Ing. Horst Zimmermann
Institute for Electrical Measurements and Circuit Design, Vienna University of Technology
Gusshausstr. 25/354, A-1040 Wien, Austria
E-Mail: kerstin.schneider@tuwien.ac.at, horst.zimmermann@ieee.org

Series Editors:

Dr. Kiyoo Itoh
Hitachi Ltd., Central Research Laboratory, 1-280 Higashi-Koigakubo
Kokubunji-shi, Tokyo 185-8601, Japan

Professor Thomas Lee
Stanford University, Department of Electrical Engineering, 420 Via Palou Mall, CIS-205
Stanford, CA 94305-4070, USA

Professor Takayasu Sakurai
Center for Collaborative Research, University of Tokyo, 7-22-1 Roppongi
Minato-ku, Tokyo 106-8558, Japan

Professor Willy M. C. Sansen
Katholieke Universiteit Leuven, ESAT-MICAS, Kasteelpark Arenberg 10
3001 Leuven, Belgium

Professor Doris Schmitt-Landsiedel
Technische Universität München, Lehrstuhl für Technische Elektronik
Theresienstrasse 90, Gebäude N3, 80290 München, Germany

ISSN 1437-0387

ISBN-10 3-540-29613-1 Springer Berlin Heidelberg New York
ISBN-13 978-3-540-29613-3 Springer Berlin Heidelberg New York

Library of Congress Control Number: 2006926218

Springer is a part of Springer Science+Business Media.

springer.com

Printed in The Netherlands

Camera-ready by the Author and SPi using a Springer LaTeX macro package
Cover concept by eStudio Calmar Steinen using a background picture from Photo Studio "SONO".
Courtesy of Mr. Yukio Sono, 3-18-4 Uchi-Kanda, Chiyoda-ku, Tokyo
Cover design: *design & production* GmbH, Heidelberg

Printed on acid-free paper SPIN: 11010838 57/3100/SPI - 5 4 3 2 1 0

Publisher Services, Pondicherry

Preface

The growing demand for high-speed, broadband data communication motivates the development of low-cost, high-performance optical receivers for fiber-optical networks. This book sets its focus especially on highly sensitive receivers with medium and high speed capability for the "last mile" connection in fiber-to-the-home (FTTH) systems. These connections are normally realized with infrared light with wavelengths of 1310 and 1540 nm. This fact makes it necessary for silicon optical receivers to use an external Ge or III/V-semiconductor based photodiode. Therefore this book deals with optical receivers for detection of infrared light, including all the problems emerging from an external photodiode, such as very high input-node capacitance somewhere in the order of pF, compared to an integrated photodiode where the input-node capacitance is about an order of magnitude less, problems due to bond-wire parasitics at the input-node, etc. The influence of these problems can be clearly seen in the performance of optical receivers. The high input-node capacitance, for example, strongly influences the bandwidth and the sensitivity.

Compared to the book *Integrated Silicon Optoelectronics* of one of the authors, which concentrates on physics and integration of photodetectors in modern silicon bipolar, CMOS and BiCMOS processes, descriptions of fabrication technologies and properties of integrated photodetectors, and *Silicon Optoelectronic Integrated Circuits*, which goes deeper into the details of the circuit design of ICs with integrated photodiodes for a wide variety of applications, this book concentrates on circuit design for optical receivers with external photodiodes for optical communication. The main subject of *Highly Sensitive Optical Receivers* is the description of the state-of-the-art of low-noise silicon amplifiers and the comparison of bipolar, CMOS, BiCMOS, as well as SiGe amplifiers.

This new book is a summary of fundamental theory and a presentation of state-of-the-art optical receiver circuits and designs. Recent optical receivers developed by the authors show the rapid progress in optical receiver design.

The first chapter explains the motivation why all optical receivers designed by the authors are done in deep-sub-micron CMOS technology.

Although these deep-sub-micron CMOS technologies cause a lot of problems, due to low power supply voltage, low Early voltage and so on, this book will show that these technologies are attractive and interesting for low-noise optical receivers for medium and high data rate applications.

In particular, the newest deep-sub-micron CMOS low-noise amplifier topologies are described in detail addressing the challenging application in optical burst-mode receivers. Thereby the excellent noise properties of deep-sub-micron CMOS receivers and fast gain switching capability are highlighted. A new approach for solving the stability problem resulting from gain switching is described. This book shows how to solve the difficulties in circuit design with deep-sub-micron CMOS technologies and how to use the benefits of the technology as for example the possibility to easily integrate the analog and the digital part in systems-on-chip (SoCs). Using a standard digital deep-sub-micron CMOS process for analog design has the disadvantage of high device tolerances to deal with, but avoids costs for technology development for analog process extensions.

In the beginning of the book in Chap. 1 the motivation for burst-mode communication and the incentive to use systems-on-chip in deep-sub-micron CMOS technology are discussed.

In Chap. 2 different kinds of networks are described. Furthermore continuous-mode and burst-mode access are compared. The additional requirements for burst-mode optical receivers will be discussed and the advantages of time-division-multiplex access (TDMA) will be pointed out. The increasing importance of burst-mode receivers is reflected in the growing amount of publications on this topic. In the beginning of the 1990s the first papers on burst-mode receivers were published. The number rapidly increased in the following years and is still growing. In Chap. 2, fundamental parts of optical receiver front-ends are also described. An essential part of optical receivers are the photodetectors. Photodetectors and especially SiGe photodetectors, therefore, are discussed in Chap. 3. The main focus of attention is on the preamplifier, being usually a transimpedance amplifier, in Chap. 4. Nevertheless, also main and limiting amplifiers are discussed.

Chapter 5 gives a short overview of an SiGe heterojunction bipolar technology, as well as some more details about the deep-sub-micron CMOS processes used for the designs described in Chap. 9.

AC-analysis as well as stability analysis of several designs are contained in Chap. 6. After the feedback theory a transimpedance amplifier with an ideal amplifier is described. This is followed by an analysis of realized circuits.

Afterwards, in Chap. 7, integrated circuit technologies of current interest are described. BiCMOS, SiGe heterojunction bipolar, submicron CMOS and deep-sub-micron CMOS technologies are compared and the advantages and disadvantages of each concerning noise are described. The device properties are compared to the properties of ideal devices and the effects of down-scaling technologies are described.

In Chap. 8 an overview of the state of the art of BiCMOS, SiGe heterojunction-bipolar and CMOS optical receivers in the literature is given. Chapter 9 summarizes the simulation environment and component models for circuit design and describes the measurement set-up and the circuits as well as printed circuit boards for characterization. Afterwards the circuits and properties of several advanced optical CMOS receivers and optical burst-mode receivers designed at the Institute for Electrical Measurements and Circuit Design at Vienna University of Technology in 0.18 μm and 0.12 μm standard digital CMOS are described in detail. Finally a summary of the characterized performance of the optical receivers is done. A comparison of the different designs and their results for optical receivers known from the literature follows.

The authors would like to thank their colleagues at the Institute for Electrical Measurements and Circuit Design at Vienna University of Technology for fruitful discussion and their valuable support, especially Franz Schlögl, Robert Swoboda, Michael Förtsch, Jürgen Leeb and Alexander Nemecek as well as the head of the institute, Gottfried Magerl, for his great support towards a quick start of research. Furthermore special thanks are directed to A. Wiesbauer, J. Hauptmann, M. Haas, and A. Martin from Infineon Technologies DC Villach and DC Vienna for their constant financial and technical support as well as the opportunity to use the design environment.

Vienna, June 2006

Kerstin Schneider
Horst Zimmermann

Furthermore, I want to thank my parents, Bernd and Petra, for their support during my studies and especially Rainer for their patience and understanding.

Vienna, June 2006

Kerstin Schneider

Contents

1 **Introduction** 1

2 **Fundamentals** 5
2.1 Point-to-Point Versus Point-to-Multipoint Access 5
2.1.1 Point-to-Point Access 5
2.1.2 Point-to-Multipoint Access 6
2.2 Continuous-Mode Versus Burst-Mode Communication 7
2.2.1 Continuous-Mode Communication 7
2.2.2 Burst-Mode Communication 7
2.3 Components of the Optical Receiver Front-End 8
2.4 Optical Fibers 9

3 **Photodetectors** 13
3.1 Basics of Photodetectors 13
3.2 Avalanche Photodiode 18
3.3 Pin Photodiode 18
3.4 SiGe Photodetectors 19
3.4.1 Heteroepitaxial Growth 19
3.4.2 Absorption Coefficient of SiGe Alloys 21
3.4.3 Ge-on-Si IR Photodetectors 21
3.4.4 SiGeC 31
3.4.5 SiGe/Si pin Hetero Bipolar Transistor Integration 32

4 **Amplifiers** 35
4.1 Preamplifier, Transimpedance Amplifier 35
4.2 Main Amplifier, Limiting Amplifier 36
4.2.1 Limiting Amplifier 37
4.2.2 AGC Amplifier 37

5 Integrated Circuit Technology ... 43
5.1 SiGe Heterojunction Bipolar (HBT) ... 43
5.2 Deep-Sub-Micron CMOS ... 45

6 Transimpedance Amplifier Theory ... 51
6.1 Feedback Theory ... 51
6.1.1 Shunt–Shunt Feedback ... 51
6.1.2 Input and Output Resistance ... 55
6.2 TIA with Ideal Amplifier ... 57
6.3 TIA with Frequency-Dependent Open-Loop Gain ... 58
6.3.1 TIA with Folded-Cascode Amplifier Stage ... 59
6.3.2 TIA with Inverter Amplifier Stages ... 63
6.3.3 Transimpedance-Gain Switching and Stability of TIA with Inverter Amplifier Stages ... 70
6.3.4 Three-Stage Burst-Mode TIA with Internal Feedback ... 76
6.3.5 Transimpedance-Gain Switching and Stability of Three-Stage Burst-Mode TIA with Internal Feedback ... 79

7 Noise Theory ... 85
7.1 Sensitivity and Power Penalty ... 85
7.1.1 Bit-Error Rate ... 85
7.1.2 Sensitivity ... 89
7.1.3 Power Penalty ... 89
7.2 Noise Models of Components ... 91
7.2.1 Resistor Noise Model ... 92
7.2.2 Bipolar- and Heterojunction-Bipolar-Transistor Noise Model ... 93
7.2.3 Field-Effect-Transistor Noise Model ... 96
7.3 Noise Models of Transimpedance Amplifier ... 99
7.3.1 Ideal-Amplifier TIA ... 99
7.3.2 TIA with Bipolar and Heterojunction Bipolar Input Stage ... 100
7.3.3 TIA with MOS Input Stage ... 100
7.3.4 Comparison of Bipolar and Field-Effect Transistor Circuits Based on Noise Theory ... 101
7.4 Noise Models of More Complex TIAs in Deep-Sub-micron CMOS Technology ... 104
7.4.1 Noise Analysis of Folded-Cascode TIA ... 104
7.4.2 TIA with CMOS-Inverter Input Circuit ... 109

8 State of the Art ... 115
8.1 Silicon Bipolar and BiCMOS Optical Receivers ... 115
8.2 SiGe Heterojunction Bipolar and SiGe BiCMOS Optical Receivers ... 121
8.3 Silicon CMOS Optical Receivers ... 126

8.4 Summary of Results of State-of-the-Art Optical Receivers 135

9 Deep-Sub-μm CMOS Circuits 139
9.1 Simulation Environment and Component Models 139
9.1.1 Simulation Environment 139
9.1.2 Photodiode Model 139
9.1.3 MOSFET Model 141
9.2 Characterization Setup 145
9.3 Designs and Properties of Optical Receivers 146
9.3.1 Folded-Cascode Transimpedance Amplifier 148
9.3.2 Three-Inverter Transimpedance Amplifier 151
9.3.3 Three-Stage Transimpedance Amplifier 154
9.3.4 Three-Stage Burst-Mode Transimpedance Amplifier with Internal Feedback 159
9.3.5 Three-Stage Burst-Mode Transimpedance Amplifier for 2.5 $\mathrm{Gb\,s^{-1}}$ 163
9.3.6 Three-Stage Burst-Mode Transimpedance Amplifier for Ultra-Fast Gain Switching 169
9.4 Summary and Comparison 176

References 183

Index 193

List of Symbols

Symbol	Description	Units
A	Area	mm^2
A_0	Low-frequency open-loop gain	
$A(f)$	Frequency-dependent gain	
A_{loop}	Loop gain	
$A_{3I}(f)$	Frequency-dependent gain of three-inverter amplifier	
$A_{FC}(f)$	Frequency-dependent gain of folded-cascode circuit	
A_{TIA}	Effective transimpedance of TIA	Ω
c	Speed of light in a medium	$cm\,s^{-1}$
c_0	Speed of light in vacuum	$cm\,s^{-1}$
C_F	Feedback capacitance	F
C_{gd}	Gate–drain capacitance	F
C_{gs}	Gate–source capacitance	F
C_j	Junction capacitance	F
C_{ox}	Gate oxide capacitance per unit area	$fF/\mu m^2$
C_{pd}	Photodiode capacitance	F
C_{para}	Parasitic capacitance	F
C_T	Input-node capacitance	F
d_{ARC}	Thickness of antireflecting coating	µm
d_I	Thickness of intrinsic region	µm
d_p	Thickness of p-type region	µm
F	Lowering factor for stability analysis	
f	Frequency	Hz
$f_{-3\,dB}$	-3 dB cut-off frequency	Hz
f_β	-3 dB cut-off frequency of β	Hz
f_g	-3 dB frequency bandwidth	Hz

Symbol	Description	Units
g_{ds}	Transistor output conductance	$\mathrm{A\,V^{-1}}$
g_{m}	Transconductance	$\mathrm{A\,V^{-1}}$
$g_{\mathrm{m}}^{\mathrm{I}}$	Transconductance of inverter	$\mathrm{A\,V^{-1}}$
h	Plack constant	Js
$\hbar$	$h/2\pi$	Js
$\langle i_0 \rangle$	Mean photocurrent for logical zero	A
$\langle i_1 \rangle$	Mean photocurrent for logical one	A
i_{L}	Leakage current of photodiode	A
$\overline{i^2}$	Spectral noise current density	$\mathrm{A^2}$
$\overline{i_{\mathrm{o}}^2}$	Spectral noise output noise current density	$\mathrm{A^2\,Hz^{-1}}$
$\overline{i_{\mathrm{b}}^2}$	Base current noise source	$\mathrm{A^2}$
$\overline{i_{\mathrm{c}}^2}$	Collector current noise source	$\mathrm{A^2}$
$\overline{i_{\mathrm{d}}^2}$	Drain current noise source	$\mathrm{A^2}$
$\overline{i_{\mathrm{g}}^2}$	Gate current noise source	$\mathrm{A^2}$
$i_{\mathrm{c}}x$	Small-signal capacitance current of stage x	A
i_{o}	Small-signal output current	A
i_{rx}	Small-signal resistance current of stage x	A
$\overline{i_{\mathrm{n,R}}^2}$	Spectral resistor noise current density	$\mathrm{A^2\,Hz^{-1}}$
$\overline{i_{\mathrm{n,in}}^2}$	Equivalent input noise current density	$\mathrm{A^2\,Hz^{-1}}$
$\overline{i_{\mathrm{n,amp}}^2}$	Equivalent input noise current density of amplifier	$\mathrm{A^2\,Hz^{-1}}$
$\overline{i_{\mathrm{n,Ti}}^2}$	Equivalent input noise current density of transistor	$\mathrm{A^2\,Hz^{-1}}$
I	Current	A
I_{in}	Input current	A
I_{ph}	Photocurrent	A
I_{B}	Base current	A
I_{C}	Collector current	A
I_{D}	Drain current	A
I_{E}	Emitter current	A
I_{S}	Source current	A
k_{B}	Boltzmann constant	$\mathrm{J\,K^{-1}}$
$k_{\mathrm{B}}T$	Thermal energy	eV
L	Length	μm
L_{B}	Inductance of bond wire	H
L_{D}	Diffusion length	nm
$\overline{n}$	Refractive index	
$\overline{n_1}$	Refractive index of fiber core	
$\overline{n_2}$	Refractive index of fiber cladding	
$\overline{n_{\mathrm{ARC}}}$	Refractive index of antireflecting coating	
$\overline{n_{\mathrm{s}}}$	Refractive index of surroundings	
$\overline{n_{\mathrm{sc}}}$	Refractive index of semiconductor	

Symbol	Description	Units
P_{char}	Characteristic polynomial	
P_{opt}	Incident optical power	W
$p_0(i)$	Probability density for a logical zero	
$p_1(i)$	Probability density for a logical one	
$\overline{P_{\mathrm{opt}}}$	Average incident optical power	W
q	Magnitude of electronic charge	As
$\overline{R}$	Reflectivity	
r_{b}	Base series resistance	Ω
r_{c}	Small-signal collector series resistance	Ω
r_{o}	Small-signal output resistance	Ω
r_{d}	Small-signal drain series resistance	Ω
r_{s}	Small-signal source series resistance	Ω
r_{DS}	Small-signal output resistance	Ω
R	Responsivity	$\mathrm{A\,W^{-1}}$
R_{F}	Feedback resistance	Ω
R_{S}	Series resistance	Ω
R_{L}	Load resistance	Ω
S	Spacing	µm
t	Time	s
t_{f}	Fall time	s
t_{r}	Rise time	s
T	Absolute temperature	K
T_{p}	Period interval	s
$\overline{v_{\mathrm{b}}^2}$	Base resistor noise voltage	$\mathrm{V^2}$
$\overline{v_{\mathrm{R}}^2}$	Spectral resistor noise voltage density	$\mathrm{V^2\,Hz^{-1}}$
$\overline{v_{\mathrm{n,amp}}^2}$	Equivalent input noise voltage density of amplifier	$\mathrm{V^2\,Hz^{-1}}$
$\overline{v_{\mathrm{n,in}}^2}$	Equivalent input noise voltage density	$\mathrm{V^2\,Hz^{-1}}$
$\overline{v_{\mathrm{n,Ti}}^2}$	Equivalent input noise voltage density of transistor	$\mathrm{V^2\,Hz^{-1}}$
V	Voltage	V
V_{BE}	Base–emitter voltage	V
V_{br}	Breakdown voltage	V
V_{DS}	Drain–source voltage	V
V_{Early}	Early voltage	V
V_{GS}	Gate–source voltage	V
V_{pin}	Photodiode voltage	V
V_{T}	Thermal voltage $k_{\mathrm{B}}\mathrm{T}\,\mathrm{q}^{-1}$	V
V_{Th}	Threshold voltage	V
V_{o}	Output voltage	V
W	Width	µm
Z_{F}	Feedback impedance of transimpedance amplifier	Ω

Symbol	Description	Units
α	Absorption coefficient	μm^{-1}
β	Current gain of bipolar transistor	
ϵ_0	Permittivity in vacuum	$F\,cm^{-1}$
ϵ_r	Relative permittivity	
ϵ_s	Semiconductor permittivity	$F\,cm^{-1}$
η	Quantum efficiency	%
η_e	External (total) quantum efficiency	%
η_i	Internal quantum efficiency	%
$\overline{\kappa}$	Extinction coefficient	
λ	Wavelength in a medium	nm
λ_0	Wavelength in vacuum	nm
ω	Angular frequency	s^{-1}
ω_g	$2\pi f_g$	s^{-1}
Γ_F	Gamma factor (2/3 for Si FETs in saturation region)	

1

Introduction

For long-haul and ultra-high-speed data communication, fiber-optical networks with optical amplifiers have become the main technology and do not require receivers with highest sensitivity [1,2]. The growing demand on broadband Internet access, however, has motivated development of low-cost, high-sensitivity optical receivers with a wide dynamic range for the optical input power. The gap of transmission bandwidth at the "last mile" can be closed by fibers-to-the-home (FTTH) at medium data rates. Figure 1.1 shows the principle of a passive optical network. The three homes in the example are connected to the post office on the right-hand side via a passive optical star coupler. The homes are sending in time-division multiplex access (TDMA) in a well-defined order. Due to the different distances between the homes and the receiver in the post office the attenuation is different and therefore the received signals are in a wide optical power range. A burst-mode receiver is necessary to handle the incoming signals.

Bipolar, CMOS, BiCMOS, and SiGe receivers are compared in this book. Therefore also the technologies are compared concerning noise, and receivers in these technologies are presented and compared. Bipolar transistors are faster than CMOS transistors with the same structure size and have the advantage of higher transconductance. The matching of bipolar transistors is also better than that of MOSFETs.

CMOS technologies have the advantage that they are faster on the market than bipolar or BiCMOS technologies. Especially BiCMOS technologies with the same minimum structure as CMOS technologies are normally available a few years later and are more expensive than CMOS processes of the same structure size. However, analog circuits in a 120 nm CMOS process can be realized with a similar or equal performance as in a sub-micron bipolar or BiCMOS process (e.g., 0.6 or 0.35 μm). If systems-on-chip including large digital parts with a large number of transistors are needed, deep-sub-micron CMOS, however, is an advantage compared to sub-micron BiCMOS technologies, because the chip area for the digital part is much less in deep-sub-micron

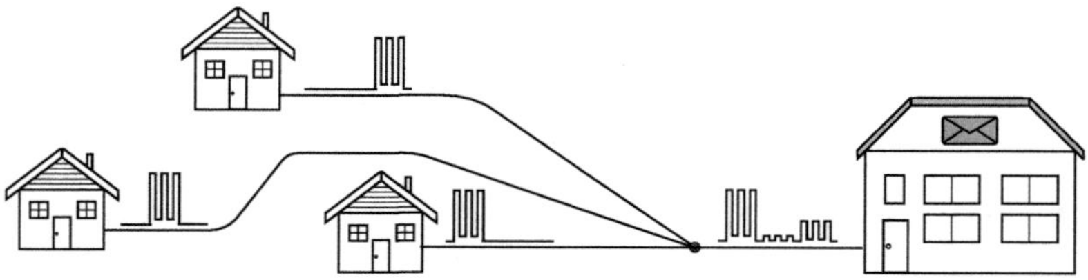

Fig. 1.1. Example configuration for a passive optical network

CMOS. It has to be mentioned, however, that for low volumes of ICs or ASICs deep-sub-micron CMOS circuits are much more expensive than sub-micron BiCMOS chips due to the large difference of mask costs.

Due to increasing doping levels of wells, channels, and substrate in down-scaling technologies, the width of the space-charge region is reduced. Also the electric field strength increases and therefore low supply voltages are necessary to stay below the breakdown field strength. These low supply voltages are the reason why classical methods of circuit design, e.g., the classical cascode circuit are no longer useable.

The low-noise optical receivers presented in this book are designed in standard digital 180 nm CMOS as well as 120 nm CMOS technology to show the low-noise capability of deep-sub-micron CMOS for high photodiode capacitance. The reason for choosing deep-sub-micron CMOS technology was the possibility to easily integrate a signal-processing digital part. The signal does not leave the chip after the optical receiver and therefore output drivers and output impedance matching can be saved in systems-on-a-chip (SoCs). Packaging costs and influences of parasitic elements, e.g., from electrostatic discharge (ESD)-structures, bond-pads and bond wires are avoided in SoCs.

To detect the infrared light of $\lambda = 1.3\,\mu\text{m}$, an external quaternary photodiode is necessary. This external photodiode causes a high input-node capacitance. The transimpedance amplifiers (TIA) in deep-sub-micron CMOS presented in this work, nevertheless, show a high sensitivity and a wide input-current range. To avoid overdrive, the transimpedance has to be variable. Therefore, usually switching of the compensation capacitance is necessary to guarantee stability. The parasitic capacitances of the MOSFET switches, however, reduce the data rate and sensitivity in these designs. Another solution for this problem is suggested here. Reducing the open-loop gain of the amplifier in order to achieve stability instead of switching the feedback capacitance enables a high dynamic optical input power range and high data rates.

The focus of this work is on the development of the preamplifier of the optical deep-sub-micron CMOS receiver. This is normally a TIA and due to the fact that there is no digital part integrated in the test-chips introduced here the biasing voltages are generated externally. Furthermore, a $50\,\Omega$ driver is implemented to drive the measurement system. In the final design the optical receiver front-end is followed by a main amplifier and the digital part and therefore the circuit for characterization is not necessary in the final system.

Fundamentals of network access and the difference between continuous-mode and burst-mode communication are discussed in Chap. 2. Furthermore the components of the optical receiver front-end are summarized. Photodetectors are described in Sect. 3. An overview of the basic knowledge about transimpedance and main and limiting amplifiers is given in Sect. 4.

In Chap. 5 a short overview of the used deep-sub-micron CMOS processes is presented. The disadvantages and the challenge of using a standard digital technology for an analog design is pointed out.

Chapter 6 processes the circuit theory of the transimpedance amplifier (TIA) with a section about stability. Chapter 7 introduces bit-error rate and sensitivity and in the following the noise theory of transistors on the one hand and TIAs on the other. First the circuit of an ideal TIA is analyzed and after this a real TIA is examined more closely. The noise theory first describes the main parameters having influence on the sensitivity, the bit-error ratio, the power penalty and the noise of the input circuit. The noise model of field-effect transistors is summarized and the noise of the input circuit is calculated for a TIA with an ideal, but noisy amplifier and for TIAs with two different CMOS input stages realized during this work.

Before our own designs are presented, an overview of the state of the art in bipolar, CMOS, BiCMOS, and SiGe optical receivers is given in Chap. 8. It includes highly sensitive optical continuous-mode receivers as well as real burst-mode receivers.

The developed designs are presented in the last chapter. First, the design environment and the measurement setup are discussed and the schematics as well as layout plots and measured results are presented. At the end of the chapter a conclusion and a comparison of our own designs with the state of the art is given.

The sensitivities achieved in 120 nm CMOS with various designs are -31.4 dBm at 622 Mb s^{-1}, -28.6 dBm at 1.25 Gb s^{-1}, and -20.4 dBm at 2.5 Gb s^{-1}, respectively. The minimum achieved switching time between minimum and maximum optical input power for burst-mode application is 1.7 ns. The maximum optical input power range is 27 dB at 1.25 Gb s^{-1} which is more than the dynamic range presented in the literature [3–6] with 21 dB. These 21 dB optical input power ranges mean that for a typical attenuation of a silica glass single-mode fiber of 0.4 dB km^{-1} at the used wavelength of 1.3 μm [7] the developed receivers are able to handle homes with distances varying by about 50 km, assuming no additional losses and dispersion.

2

Fundamentals

This chapter describes the main fundamentals of network access, compares point-to-point and multipoint access in Sects. 2.1.1 and 2.1.2, and burst-mode (BM) versus continuous-mode (CM) communication are discussed in Sect. 2.2. At the end of the chapter an overview of the components of optical transceivers is given in Sect. 2.3.

2.1 Point-to-Point Versus Point-to-Multipoint Access

There are two basic types of network links. On the one hand there is the simple point-to-point connection and on the other hand the point-to-multipoint access.

2.1.1 Point-to-Point Access

Figure 2.1 shows a typical point-to-point connection between two central offices (CO). This may be an SONET OC-192 link operating at $10\,\text{Gb}\,\text{s}^{-1}$. These connections are in nearly every range of bit rates and distances, from undersea light wave systems to very short chip interconnects [7].

More complex networks can be built by combining point-to-point connections to ring systems or active star networks. The active star is built by a hub converging several network links, for example, in Gigabit Ethernet. The active star, e.g., the hub, contains receivers and transmitters for every network link, and therefore the network consists of point-to-point connections. A different kind of connection is with a passive optical star coupler, where the coupling is done by a passive optical device. The standard transmission of point-to-point networks is CM (see Sect. 2.2.1) communication. The communication can be unidirectional or bidirectional, for example in space division multiplexing (SDM), where two fibers are used in parallel, one for uplink, one for downlink communication, or in wavelength division muliplexing (WDM),

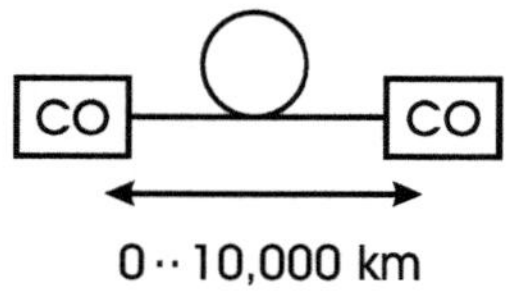

Fig. 2.1. Point-to-point connection

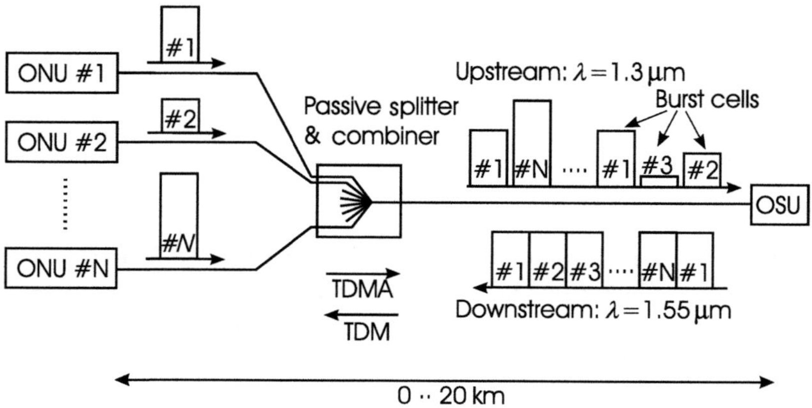

Fig. 2.2. Point-to-multipoint network [8]

where only one fiber is used, but uplink and downlink communication is done in different wavelengths. If only one fiber and wavelength is available, bidirectional communication can be realized by BM transmission (see Sect. 2.2.2).

2.1.2 Point-to-Multipoint Access

Point-to-multipoint networks are so-called passive optical networks (PON) like that illustrated in Fig. 2.2. They are used as fiber-to-the-home (FTTH) systems where several optical network units (ONU), for homes, for example, are connected to an optical line terminal (OLT) via fibers with a passive optical star coupler. PONs are limited to relatively short distances of about 20 km at maximum and are currently operated at bit rates of about 50–155 $\mathrm{Mb\,s^{-1}}$ [7].

ATM-PON (asynchronous transfer mode) is the most common PON system and is defined in the full services access network (FSAN) standard. It is also defined in ethernet passive optical network (EPON). The typical ATM-PON is built by about 16–32 homes or ONUs which share a data rate of, e.g., 155 $\mathrm{Mb\,s^{-1}}$, which means that every home has a time slot inversely proportional to the number of ONUs with the maximum bit rate. To avoid data collision the ONUs send so-called data bursts in these time slots. Each ONU is allowed to send in a certain order defined in a protocol. This leads to an average data rate of about 5–10 $\mathrm{Mb\,s^{-1}}$ for each ONU. This can be used for Internet access, TV service, and telephone. This type of sharing the optical medium is so-called time division multiplex access (TDMA). The TDMA is used for upstream communication in PON. The downstream communication is realized in a different wavelength, e.g., $\lambda = 1{,}300$ nm for upstream communication and $\lambda = 1{,}550$ nm for downstream communication. In this scenario the downstream communication is realized by CM communication. The OLT sends in CM to all ONUs, which share the data in time-division multiplex (TDM), which means that every ONU selects the information with the suitable address.

2.2 Continuous-Mode Versus Burst-Mode Communication

There are two different types of transmission modes: CM and BM. The signals corresponding to these two modes are shown schematically in Fig. 2.3.

2.2.1 Continuous-Mode Communication

In CM transmission a steady, uninterrupted stream of bits is transmitted (see Fig. 2.3a). Observed over a long time period the signal shows the same number of ones and zeros, which means that the signal is dc balanced. This means that the average value is centered in the middle between one and zero. This quality offers the possibility of ac coupling between the individual circuit blocks [7], if this time period is not too long.

2.2.2 Burst-Mode Communication

In BM data transmission, different senders transmit short bursts. Between two bursts there is a short time slot of silence. Figure 2.3b shows the principle of a BM data signal. The bursts are normally much longer than displayed in Fig. 2.3b; they are about 400 bits long [7]. The different bursts are normally differently attenuated and therefore show strongly different optical power. The different bursts also show strongly different average levels. The time slot between the bursts gives the BM receiver the possibility for gain setting and settling to enable the correct handling of a wide range of optical input power. The actual information is headed by some preamble bits for synchronization of the clock, due to the fact that the data bursts are asynchronous.

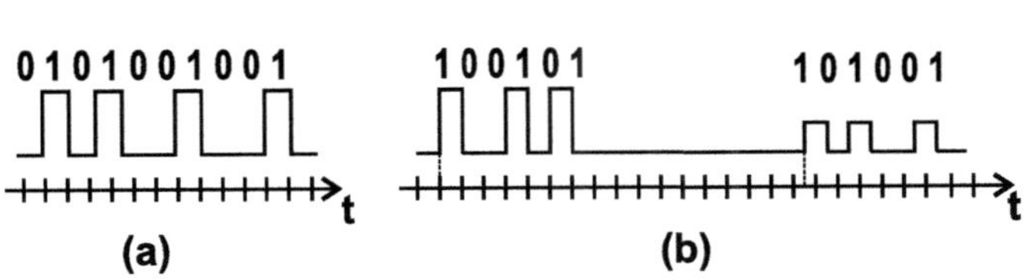

Fig. 2.3. Principal types of data communication: (**a**) CM data; (**b**) BM data

The minimal and maximal values of the BM signal vary, depending on the distance from sender to receiver and on the burst activity. High activity leads to an average value halfway between high and low; for low activity the average value is decreasing towards zero. This behavior makes dc coupling necessary. The lack of ac-balancing and the different arriving optical powers from different senders make it necessary to design special receivers for BM transmission. In BM, the period between two bursts can be shorter than depicted in Fig. 2.3.

According to the global standard G.983.1 class C announced by ITU-T in 1998, the burst cells at an OLT can have a very large power difference [8]. In addition, the laser diodes are biased above threshold to achieve a high data rate and a low jitter. The receiver in an OLT, therefore, must realize high sensitivity, wide dynamic range (more than 30 dB difference between strong and weak cells), and it must receive signals with an extinction ratio as low as 10 dB.

2.3 Components of the Optical Receiver Front-End

A block diagram of a transceiver front-end is shown in Fig. 2.4. The receiver part starts with the PD (PD) which receives the optical signal from the fiber. The photocurrent is amplified and converted to a voltage by the preamplifier, which is usually realized by a transimpedance amplifier (TIA). This voltage is further amplified by a limiting amplifier (LA) or an automatic gain control (AGC) amplifier, the so-called main amplifier (MA). The output signal of the MA is high enough to be fed into a clock and data recovery circuit (CDR) where the clock is extracted and the data retimed. A demultiplexer (DMUX) converts the serial high-speed data stream into n parallel low-speed data streams which can be processed by a conventional digital logic circuit.

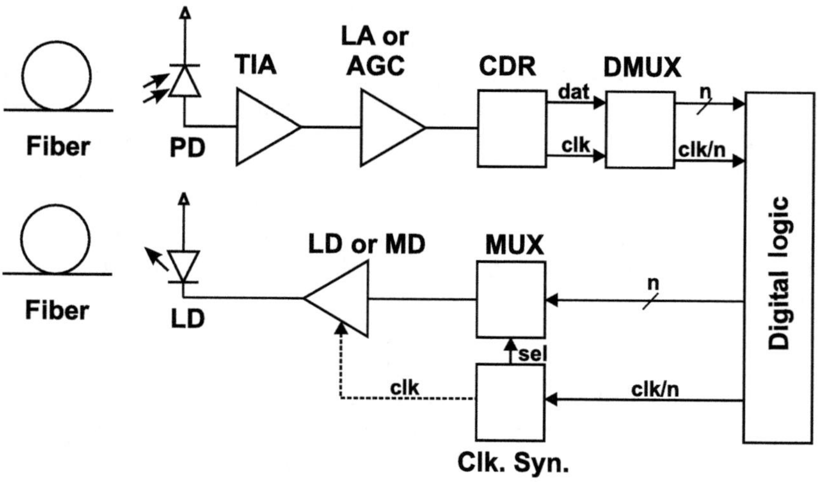

Fig. 2.4. Block diagram of transceiver front-end [7]

The transmitter processes similar steps the other way round. The data arrive as parallel low-speed data streams and are combined to a serial high-speed data stream by a multiplexer (MUX). The MUX is controlled by a clock which is synthesized to the slower word clock of the arriving data. The synthesizing happens in the clock synthesizer which also generates the clock for the laser driver (LD), or for the modulator driver (MD), respectively. This laser or MD drives the optical transmitter. In the case of a laser driver the LD modulates the current which is fed into the laser diode. The MD on the other hand works with an optical modulation of a continuous-wave laser.

In the following section, the optical and analog parts of the optical *receiver* front-end shown in Fig. 2.4 are described.

2.4 Optical Fibers

There are three basic types of optical fibers. Their main characteristics are displayed in Fig. 2.5. The three types are multimode fibers with stepped or graded index and single-mode fibers.

The *multimode fiber with stepped index* is the most simple light guide. A circular core with constant refractive index $\overline{n}_1$ is covered by a cladding material with a lower refractive index $\overline{n}_2$.

The incoming light pulse is transported through the fiber in several modes. A total internal reflection happens for all modes which meet a critical angle θ_0. These modes are guided through the fiber. All other modes will be refracted and therefore lost for the transfer.

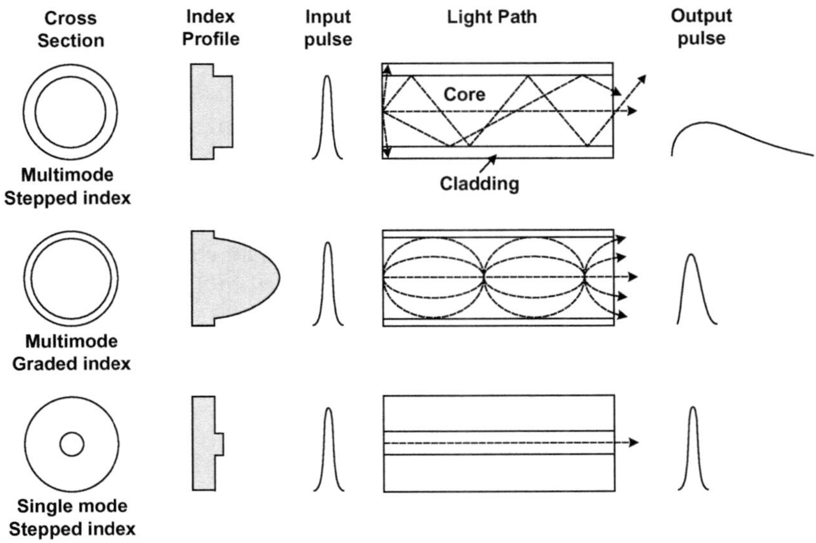

Fig. 2.5. Characteristics of three basic types of optical fibers [9]

The numerical aperture (NA) defines the maximum angle θ_0 where total internal reflection happens

$$\mathrm{NA} = \sin\theta_0 = \sqrt{\overline{n}_1^2 - \overline{n}_2^2} \approx \overline{n}_1\sqrt{2\Delta} \tag{2.1}$$

with

$$\Delta = (\overline{n}_1 - \overline{n}_2)/\overline{n}_1. \tag{2.2}$$

The number of modes which can propagate in a fiber is defined by Maxwell's electromagnetic field equations. They are related to the dimensionless quantity V,

$$V = \frac{2\pi a}{\lambda}\overline{n}_1\sqrt{2\Delta}, \tag{2.3}$$

where λ is the wavelength of the incoming light and a is the radius of the core [9, 10]. For $V \leq 2.405$ only a single mode of light can propagate.

Multimode fibers with stepped index show a strong distortion due to the different lengths of the light-paths through the fibers for different modes (see Fig. 2.5). This distortion is called intermodal dispersion. The intermodal dispersion limits the transmission bandwidths; it is therefore nearly reciprocally dependent of the length of the fiber and therefore the transmission bandwidth is normally given as bandwidth-length-product. A transmission bandwidth of about 5 MHz km are typical for step index multimode fibers [10].

To minimize the distortion of the output pulse the refractive index of the core is carried out graded. This leads to periodic light paths (see Fig. 2.5 in the middle) where in general higher modes travel longer paths than the lower modes. Due to the lower refractive index in the periphery of the core the velocity of the travelling light is higher on the long paths than in the middle, where the refractive index is maximum and the path is the shortest. This means that the transit times of all modes are nearly equal and therefore the distortion of the output pulse is minimal for *multimode fibers with graded index*. These fibers show transmission bandwidths of 0.2–3 GHz km, see Table 2.1.

The effects of pulse distortion due to different travel times of modes do not take place for *single-mode fibers.* The single-mode fiber has a small core diameter of about 4–10 μm. The change of the refraction index between core and cladding is less than for multimode fibers. The characteristic values of the optical fibers shown in Fig. 2.5 are summarized in Table 2.1.

Table 2.1. Overview of characteristic values of optical fibers [10]

	multimode fiber		single-mode fiber
	w. stepped index	w. grated index	
transmission bandwidth (GHz km)	< 0.05	0.2..3	$\gg$1
core diameter (μm)	50..600	50; 62.5	4..10
NA	0.2..0.4	$\cong$0.2	0.1..0.2
V	$\gg$2.4	$\gg$2.4	$\ll$2.4

Other sources of pulse spreading take place in single-mode fibers as well as in multimode fibers. The optical signal is attenuated as it travels through a long stretch of fiber because of scattering, material impurities and other effects. Due to the fact that the attenuation is proportional to the length of the fiber, the fiber attenuation and fiber loss is specified in $\mathrm{dB\,km^{-1}}$. The attenuation depends on the wavelength. For example silica glass has two low-absorption windows, one at the wavelength of 1.3 μm and one at 1.55 μm. A single-mode fiber has a loss of about 0.25 dB km at $\lambda = 1.55$ μm and of about $0.4\,\mathrm{dB\,km^{-1}}$ at $\lambda = 1.3$ μm [7].

Plastic optical fibers (POF) are very cheap and therefore the use of low-cost optical connectors is useful. These lead to optical receivers with large photodetectors. The loss is with about $180\,\mathrm{dB\,km^{-1}}$ at 650 nm wavelength much higher than for the silica glass fibers in their low-loss windows. The usual applications for POF are therefore short-distance applications for example in cars.

3

Photodetectors

There are different semiconductor materials that are used for photodetectors. The most common materials and the most interesting types of photodetectors as well as their basics will be introduced in this section.

3.1 Basics of Photodetectors

The first element on the receiver side of the transceiver front-end is the PD. Its main properties are quantum efficiency, speed, capacitance and leakage current. Quantum efficiency and speed are determined by the optical absorption coefficient α (see Fig. 3.1). Ge, GaAs, InP, and InGaAs have a much larger optical absorption coefficient and are therefore most interesting for photodetectors. Especially Ge and InGaAs are very appropriate for optical BM receivers at 1.3 µm light. The capacitance C_{PD}, the responsivity R and the noise as well as the noise of the following preamplifier directly influence the sensitivity of the whole receiver.

The total, external or overall *quantum efficiency* η is defined as the number of photogenerated electron–hole pairs, which contribute to the photocurrent, divided by the number of incident photons. The external quantum efficiency can be determined, when the photocurrent of a photodetector is measured for a known incident optical power.

A fraction of the incident optical power is reflected (see Fig. 3.2) due to the difference in the index of refraction between the surroundings $\bar{n}_{\mathrm{s}}$ (air: $\bar{n}_{\mathrm{s}} = 1.00$) and the semiconductor $\bar{n}_{\mathrm{sc}}$ (e.g., Si, $\bar{n}_{\mathrm{sc}} \approx 3.5$). The reflectivity $\bar{R}$ depends on the index of refraction $\bar{n}_{\mathrm{sc}}$ and on the extinction coefficient $\bar{\kappa}$ of an absorbing medium, for which the dielectric function $\bar{\epsilon} = \bar{\epsilon}_1 + \mathrm{i}\bar{\epsilon}_2 = (\bar{n}_{\mathrm{sc}} + \mathrm{i}\bar{\kappa})^2$ is valid ($\bar{n}_{\mathrm{s}} = 1$) [11]:

$$\bar{R} = \frac{(1 - \bar{n}_{\mathrm{sc}})^2 + \bar{\kappa}^2}{(1 + \bar{n}_{\mathrm{sc}})^2 + \bar{\kappa}^2}. \tag{3.1}$$

The extinction coefficient is sufficient for the description of the absorption. The absorption coefficient α can be expressed as:

$$\alpha = \frac{4\pi\bar{\kappa}}{\lambda_0}. \tag{3.2}$$

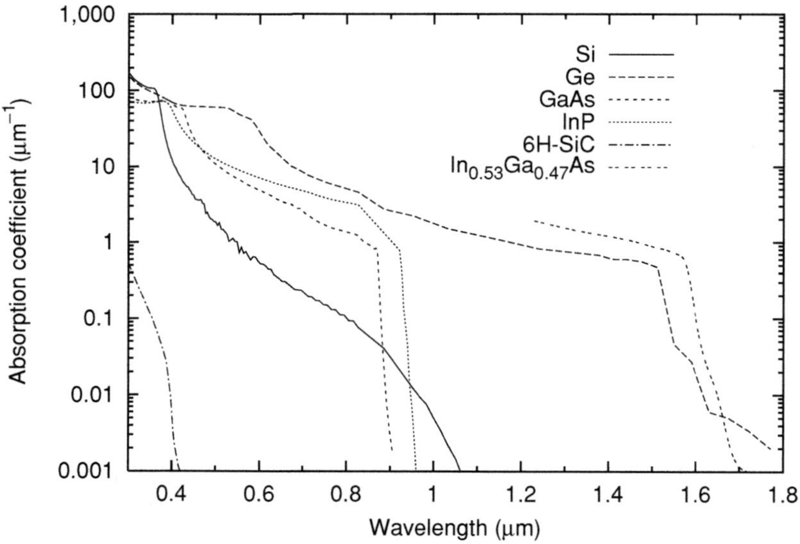

Fig. 3.1. Absorption coefficient of important semiconductor materials versus wavelength

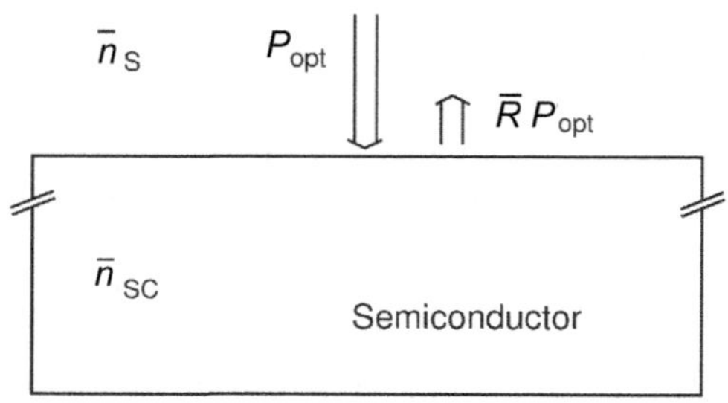

Fig. 3.2. Reflection at the semiconductor surface

The optical quantum efficiency η_o can be defined in order to consider the partial reflection:

$$\eta_o = 1 - \bar{R}. \tag{3.3}$$

The reflected fraction of the optical power can be minimized by introducing an antireflection coating (ARC) with thickness d_{ARC} (see Fig. 3.3):

$$d_{ARC} = \frac{\lambda_0}{4\bar{n}_{ARC}}. \tag{3.4}$$

The index of refraction of the ARC layer can be calculated:

$$\bar{n}_{ARC} = \sqrt{\bar{n}_s \bar{n}_{sc}}. \tag{3.5}$$

The optimum index of refraction of the ARC-layer is determined by the refractive index $\bar{n}_s$ of the surroundings and by the refractive index $\bar{n}_{sc}$ of the

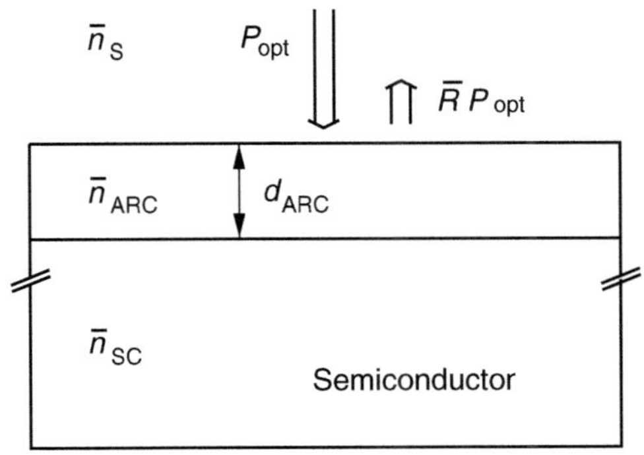

Fig. 3.3. Semiconductor with an antireflection coating

semiconductor. A complete suppression of the partial reflection, however, is not possible in practice. For silicon photodetectors, SiO_2 ($\bar{n}_{sc} = 1.45$) and Si_3N_4 ($\bar{n}_{sc} = 2.0$) ARC-layers are most appropriate.

Because of the partial reflection, it is useful to define the internal quantum efficiency η_i as the number of photogenerated electron–hole pairs, which contribute to the photocurrent, divided by the number of photons which penetrate into the semiconductor.

The external quantum efficiency is the product of the optical quantum efficiency η_o and of the internal quantum efficiency η_i:

$$\eta = \eta_o \eta_i . \tag{3.6}$$

The internal quantum efficiency η_i will be discussed later in this section after dynamical quantum efficiency has been introduced.

For the development of photoreceiver circuits, and especially of transimpedance amplifiers, it is interesting to know how large the photocurrent is for a specified power of the incident light with a certain wavelength. The *responsivity* R is a useful quantity for such a purpose:

$$R = \frac{I_{ph}}{P_{opt}} = \frac{q\lambda_0}{hc}\eta = \frac{\lambda_0 \eta}{1.243}\ \mathrm{A\,W^{-1}}, \tag{3.7}$$

where λ_0 is in μm. The responsivity is defined as the photocurrent I_{ph} divided by the incident optical power. R depends on the wavelength, therefore the wavelength has to be mentioned if a responsivity value is given.

The dashed line shown in Fig. 3.4 represents the maximum responsivity of an ideal photodetector with $\eta = 1\%$ or 100%. The responsivity of real detectors is always lower due to partial reflection of the light at the semiconductor surface and due to partial recombination of photogenerated carriers in the semiconductor or at its surface.

Strictly speaking we have to distinguish between the stationary and the dynamical internal quantum efficiency. In the stationary case, the light intensity

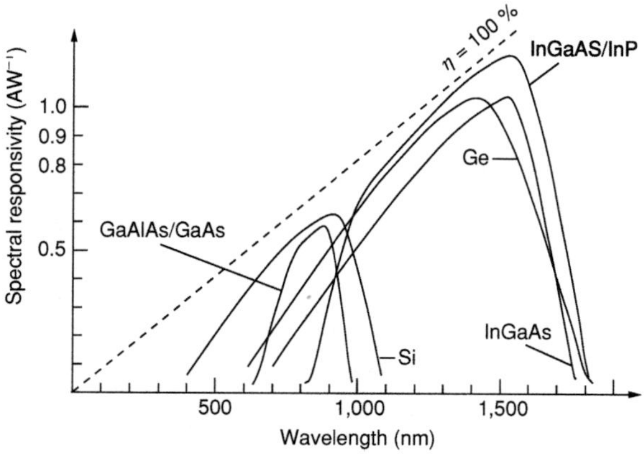

Fig. 3.4. Comparison of the responsivity of real photodetectors with an ideal photodetector with a quantum efficiency $\eta = 1$ (100%) [12]

and the photocurrent are constant with time. In the dynamical case both change with time. The dynamical quantum efficiency generally is lower than the stationary one.

Let us discuss the stationary case first. As explained above, practically all carriers which are photogenerated in drift regions contribute to the photocurrent. In other words, there is no negative influence of recombination on the internal quantum efficiency in the space-charge regions (SCRs) of PDs. The recombination of photogenerated carriers in region 1 and 2 (see Fig. 3.5), however, reduces the internal quantum efficiency. In the highly doped region 1, the carrier lifetime is reduced considerably. This reduces the internal quantum efficiency for short wavelengths considerably, because a large portion of the light is absorbed in region 1.

Light with long wavelengths penetrates deeper into semiconductors and the recombination of photogenerated carriers in region 2 can reduce the internal quantum efficiency. The recombination of photogenerated carriers in region 1 is not very important for long wavelengths due to the large penetration depth and the small portion of photogenerated carriers in region 1.

In the dynamical case, carriers being photogenerated in region 1 and especially in region 2 do not have enough time to diffuse to the space-charge or drift region before the light intensity is reduced again. The diffusion tails of the photocurrent of consecutive light pulses overlap (Fig. 3.6). The photocurrent for a sine-wave light modulation reduces similarly at high frequencies. It should be mentioned explicitly that the dynamical quantum efficiency depends on the frequency or data rate. The higher these both are, the smaller the dynamical quantum efficiency becomes until the minimum is achieved. This minimum is set by the portion of carriers being generated in the space-charge region, when we assume that the frequency is not extremely high and all drifting carriers still reach the boundary of the space-charge region and

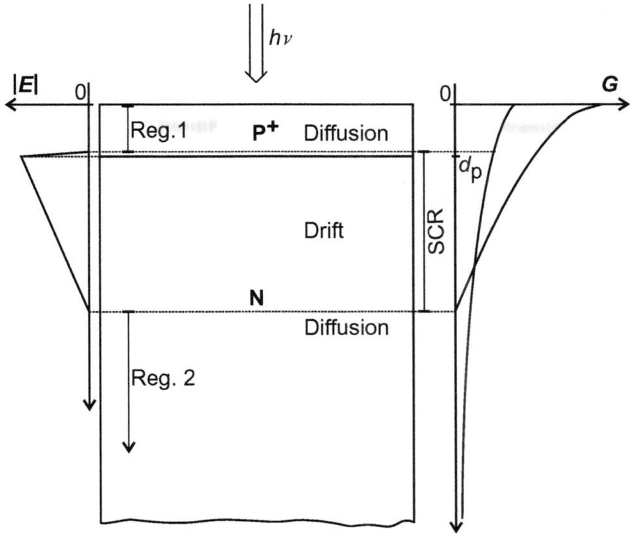

Fig. 3.5. Drift and diffusion in a PD [13]

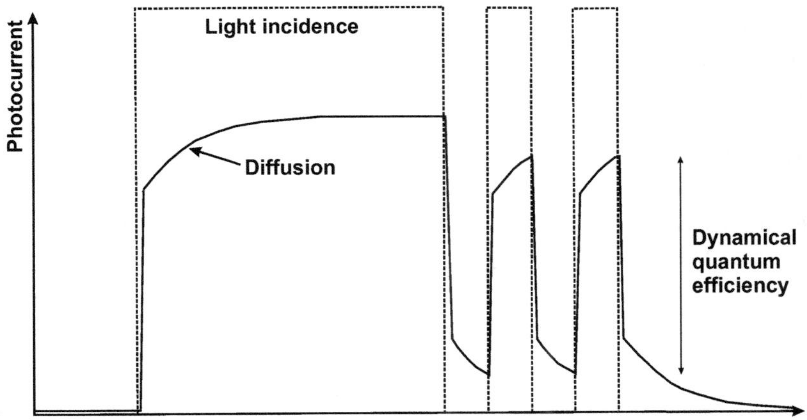

Fig. 3.6. The influence of carrier diffusion on the dynamical quantum efficiency of photodetectors at high data rates [14]

contribute to the photocurrent. For this case, we can use the expression

$$\eta_{\mathrm{i}} = (1 - \exp[-\alpha(d_{\mathrm{p}} + d_{\mathrm{I}})]) \exp(-\alpha d_{\mathrm{p}}) \tag{3.8}$$

to describe the dynamical internal quantum efficiency. The optical absorption coefficient α is most important in this expression. This expression was derived for ideal pin PDs with the thickness d_{I} of the intrinsic region, i.e., the thickness

d_I of the drift region. We can also use (3.8) for a pn PD shown in Fig. 3.5 to a good approximation because the space-charge region with the thickness d_I does not penetrate far into the highly doped P^+ layer.

3.2 Avalanche Photodiode

The avalanche photodiode (APD) is a reverse biased diode. It features four layers shown in Fig. 3.7. The multiplication region provides gain through avalanche multiplication of the electron–hole pairs generated in the intrinsic (i-) layer, the so-called absorption region. To activate the avalanche process, the APD shown in Fig. 3.7 must be operated at a fairly high reverse bias voltage of about 40–60 V or even more [7]. This type of PD is only used in three of the receivers described in the state of the art and therefore not discussed any further.

3.3 Pin Photodiode

A pin PD consists of a p–n junction with an intrinsic (undoped) layer in-between (see Fig. 3.8). The junction is reverse biased, about 3–10 V, to achieve a strong electric field in the intrinsic layer. The incoming photons create electron–hole pairs in the intrinsic layer (i-layer). These pairs are immediately separated by the high electric field and an electrical current begins to flow. The efficiency and the speed of the PD depend on the width W of the i-layer. The wider the i-layer is, the higher is the efficiency and on the other hand, the smaller W, the faster the electrons and holes move through the i-layer and therefore the speed of the pin PD rises. Therefore a tradeoff between

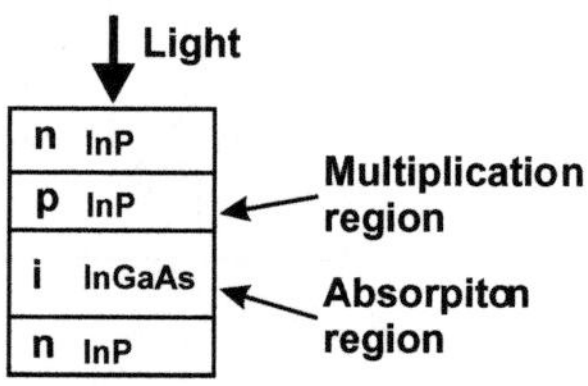

Fig. 3.7. Avalanche PD (schematic, cross section)

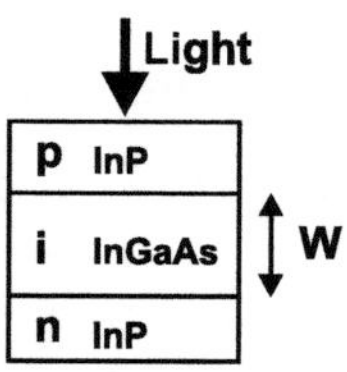

Fig. 3.8. Schematic cross section of a pin PD

efficiency and speed must be found. The absorption coefficient α is high in InGaAs pin PDs and therefore W need not be as large as in an Si pin PD to achieve a high quantum efficiency. Furthermore InGaAs has a very high electron mobility compared to Si. Both facts favor InGaAs with respect to speed. InGaAs can be used at 1.3 μm and 1.54 μm, whereas Si cannot detect both of these wavelengths.

3.4 SiGe Photodetectors

The availability of SiGe BiCMOS technologies for circuit production leads to the idea to investigate SiGe photodetectors. The higher α (see Fig. 3.9) of Ge than that of Si makes Ge or SiGe photodetectors on Si interesting.

SiGe alloys allow the integration of infrared detectors on Si. The addition of Ge to Si also increases the absorption coefficient in the spectral range from 400 to 1,000 nm, allows a reduction of the detector thickness, and, therefore, enables faster detectors than with pure Si. To exploit the advantages of SiGe alloys for Si-based photodetectors, however, the problems associated with the lattice constant mismatch and energy band discontinuities have to be understood. This section gives an overview of these aspects and describes several examples of SiGe photodetectors.

3.4.1 Heteroepitaxial Growth

The growth and physical properties of $Si_{1-x}Ge_x$ heteroepitaxial layers on Si have been investigated for more than 20 years [15–19] finally producing a technology which is entering high-volume and large-scale manufacturing of heterojunction bipolar transistors (HBTs) [20] and SiGe–HBT–BiCMOS circuits [21,22]. Optoelectronic applications of $Si_{1-x}Ge_x$/Si devices also have been developed. Medium-wavelength (1.3–1.55 μm) photodetectors for optical communication [23–25], 2–12 μm infrared photodetectors for two-dimensional

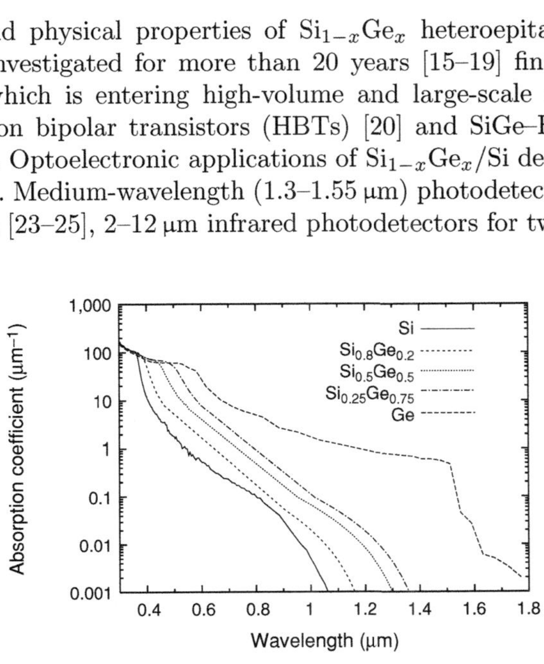

Fig. 3.9. Absorption coefficient for SiGe with different compositions [44]

focal plane arrays for thermal imaging and night vision [26–28], optical waveguides [29], and infrared light emitters for chip-to-chip optical interconnects [30, 31] have been suggested.

On the way to introducing optoelectronic devices into VLSI and ULSI to obtain a silicon-based *superchip* [32, 33], however, the following significant limitations of the SiGe/Si material system have been discovered [34]:

(i) Due to the large lattice constant of Ge, which is about 4.2% larger than that of Si, a critical thickness of $Si_{1-x}Ge_x$ layers on Si for thermal stability against misfit dislocation generation due to lattice mismatch exists [35]. This critical thickness is only about 15 nm for a Ge fraction x of 0.2 with a resulting bandgap change of approximately 150 meV [36]. Due to this low critical thickness, the quantum efficiency of normal incidence photodetectors relying on photogeneration across the $Si_{1-x}Ge_x$ bandgap is severely limited.

(ii) When an $Si_{1-x}Ge_x$ layer with a thickness greatly exceeding the critical thickness is grown directly onto Si, the layer relaxes but forms a dislocation density at the surface of order $10^{12}\,cm^{-2}$ [37]. This dislocation density reduces the carrier mobility and increases leakage currents significantly and is far too large to allow economic yields of devices in production [22]. Although relaxed $Si_{1-x}Ge_x$ buffer layers with the bulk lattice constant of the $Si_{1-x}Ge_x$ alloy on top of the Si substrate can be introduced and pseudomorphic heterostructures, which have a lateral lattice constant larger than that of Si, can then be grown on top of the buffer [38,39], these relaxed buffers still reduce the threading dislocation density only to below $10^4\,cm^{-2}$ for $Si_{0.8}Ge_{0.2}$ [37]. This density is already comparable to many III/V systems [22]. More recently, a number of annealing steps during growth have reduced the dislocation density to $10^2\,cm^{-2}$ for $Si_{0.8}Ge_{0.2}$ and $x < 0.2$ [40]. Dislocations, however, still pose a problem for higher Ge fractions and future large-scale optoelectronic applications, therefore, remain uncertain.

(iii) The dopant diffusion during thermal processing for device fabrication often degrades the nanometer precision in the placement of dopant atoms required for heterojunction devices, which can be achieved during the initial growth of heterostructures at temperatures below 700°C [41, 42].

(iv) Under all combinations of strain, $Si_{1-x}Ge_x$ is an indirect-bandgap material, resulting in inefficient emission of light and inefficient detection due to a small absorption coefficient compared to pure Ge, GaAs, or InGaAs, at least for $x < 0.75$. Despite attempts to create a direct bandgap material by the *zone-folding* principle in short-period Si/Ge superlattices [43], no breakthrough in such an approach has been achieved.

Let us continue with light absorption of SiGe alloys in Sect. 3.4.2 and with examples of infrared Ge and SiGe detectors on Si exploiting the capability of absorption in Ge and SiGe at 1.3 μm in Sect. 3.4.3. Then SiGeC for the solution of problems (iii) and (ii) will be discussed in Sect. 3.4.4.

3.4.2 Absorption Coefficient of SiGe Alloys

The exchange of Si atoms by Ge atoms increases the absorption coefficient. Furthermore the bandgap reduces with increasing Ge fraction and wavelengths longer than 1,100 nm can be detected. Figure 3.9 shows the absorption coefficient for Ge fractions of 20%, 50%, and 75%.

From Fig. 3.9 it can be concluded that SiGe detectors are not appropriate for 1.3 μm light as long as the Ge component does not considerably exceed 75%. It is advantageous to use SiGe detectors instead of Si detectors, when longer wavelengths, a thinner absorbing layer, a higher quantum efficiency and/or a higher speed of the detector are needed. In the following, several examples of SiGe receivers using these advantages will be described.

3.4.3 Ge-on-Si IR Photodetectors

The integration of Ge photodetectors on Si substrates is interesting because of the lower bandgap of Ge enabling Si-based detectors for 1.3 μm and to a somewhat less advantageous extent for 1.55 μm [45–48]. Due to the large lattice mismatch between Si and Ge of about 4%, the most effective way to fabricate high-quality SiGe and Ge layers on Si substrates is to implement graded composition buffer layers [49]. With the increase in Ge composition, however, the surface roughness generally increases. The surface roughness is caused by the influence of strain and misfit dislocations on local growth rate in SiGe mesas. Growth on miscut Si (100) substrates reduces surface roughness and dislocation densities. Thermal mismatch between the Si and Ge expansion coefficients ($\alpha_{Si} = 3.55 \times 10^{-6}\,K^{-1}$ and $\alpha_{Ge} = 7.66 \times 10^{-6}\,K^{-1}$ at 750°C), however, still leads to undesirable tensile stresses during cool-down from the growth temperature, which can form microcracks or residual tensile strain and dislocations.

High-quality Ge layers on optimized relaxed buffers (ORBs) by introducing an intermediate chemical mechanical polishing (CMP) step at $Si_{0.5}Ge_{0.5}$ in the graded structure, therefore, have been developed [50]. The CMP step liberates dislocations and creates the necessity to nucleate new dislocations. An optimized relaxation of the graded buffer results in such a way, where existing threading dislocations are more effectively used to relieve stress. Samples grown on ORB with the CMP step at 50% Ge and final pure Ge, showed a decrease in the threading dislocation density by a factor of 5–8 to $\approx 2 \times 10^6\,cm^{-2}$. This reduction in threading dislocation density led to a record low dark current density of $0.15\,mA\,cm^{-2}$ in the Ge mesa PDs [25] shown in Fig. 3.10.

The pn PDs were fabricated in Ge layers grown on an ORB. The Ge layers were in situ doped with PH_3 and B_2H_6 to obtain n- and p-type layers, respectively. The persistent PH_3 in the UHV/CVD reactor resulted in a graded pn junction. The peak p and n doping levels were $1.18 \times 10^{18}\,cm^{-3}$ and $1.3 \times 10^{18}\,cm^{-3}$, respectively. The contacts to the n-type Ge layer were

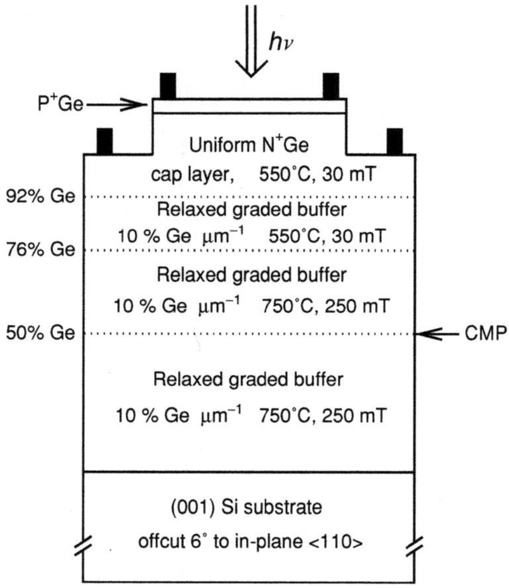

Fig. 3.10. Ge PD on SiGe/Si [25]

made by etching and patterning different-sized square mesas with sides ranging from 95 to 250 µm. The contact to the p-type Ge was structured on top of the mesas. Ti/Pt contacts were used for the n- and p-type Ge. The series resistances of the Ge diodes were 55 Ω for the 95 µm diodes and 26 Ω for the 250 µm diodes. An ideality factor for $U_{\mathrm{forward}} < 0.3\,\mathrm{V}$ of 1.1 was found. For a reverse bias of $-1\,\mathrm{V}$, the reverse current density ranged from 0.15 to $0.22\,\mathrm{mA\,cm^{-2}}$ being almost two orders of magnitude lower than in [51] for Ge diodes integrated on Si substrates. At a reverse bias of $-3\,\mathrm{V}$, the 95 µm diodes had a dark current of 0.45 µA. For the 250 µm diodes a dark current of 0.7 µA was measured. A responsivity of $0.133\,\mathrm{A\,W^{-1}}$ ($\eta_{\mathrm{e}} = 12.6\%$) was measured at the Ge PDs integrated on ORB SiGe/Si structures. These values are reasonable for Ge PDs without ARC and with a narrow depletion region of 0.24 µm because the absorption length $1/\alpha$ for $\lambda = 1.3\,\mu\mathrm{m}$ is about 1.4 µm. A better device design with ARC and pin PD structure can improve the quantum efficiency considerably. The bandwidth of 2.3 GHz has been estimated for the pn-Ge diode with the 0.24 µm depletion region [25].

One aspect should be mentioned to show the difficulty of Ge PD integration on Si. When we add the layer thicknesses of the buffer shown in Fig. 3.10, we obtain a height of 9.2 µm plus 1.5 µm for the top n^+ and p^+ layers. The metal interconnects from the Ge PD to circuits in the Si substrate, therefore, will cause severe problems.

An interdigitated Ge pin PD (Fig. 3.11) was reported in [52]. This lateral PD consists of n^+ and p^+ fingers in a 1 µm thick Ge layer on top of a 10 µm thick graded SiGe buffer layer. The finger width was 1 µm with a finger

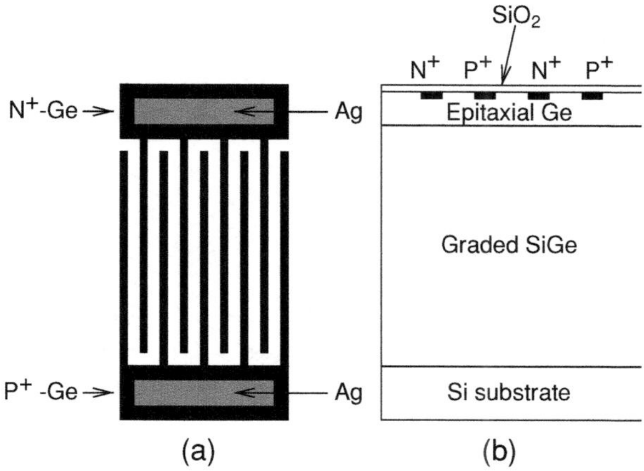

Fig. 3.11. Lateral Ge PD on thick graded SiGe buffer layer: (**a**) top view; (**b**) cross section [52]

spacing of 2 μm. The active area had a size of $25 \times 28\,\mu\text{m}^2$. With 1.3 μm light, bandwidths of 2.2, 3.5, and 3.8 GHz at reverse voltages of −1, −3, and −5 V, respectively, were measured. The external quantum efficiency was 49% at this wavelength.

A growth rate of 4.5–6.0 nm s^{-1} was achieved using low-energy plasma enhanced chemical vapor deposition. The Ge epitaxial layer had a threading dislocation density of $10^5\,\text{cm}^{-2}$ and an rms surface roughness of 3.28 nm. The dark current was 3.2 and 5.0 μA at −3 and −5 V, respectively.

Another Ge-on-Si PD was described in [53] for vertical and in-plane detection. The Ge devices are grown on a virtual substrate built by an only 31 nm thick strain-relaxed Ge buffer layer to fit the lattice constant of Ge to the one of the Si substrate. The responsivity for the vertical pin-PD was 117 mA W^{-1} with a bandwidth of 1.5 GHz with zero biasing at a wavelength of 1298 nm. The in-plane detector achieved a zero-bias responsivity of 70 mA W^{-1} with 4.4 GHz bandwidth. A bandwidth of 6.2 GHz was measured with a bias voltage of −2 V. The responsivities for 1,580 nm light were 26 and 19 mA W^{-1} for the lateral and the vertical pin-PD, respectively. The drawback of this structure is the large dark current of more than 10 μA at −1 V bias voltage. The device structure is depicted in Fig. 3.12.

A vertical Ge pin PD on Si was presented in [54]. Two Si_xGe_{1-x} buffer layers were used. Figure 3.13 shows the cross section of this Ge PD.

By optimizing the Ge concentration in the two thin SiGe buffer layers, the dislocation density in the Ge layer can be reduced by trapping many threading dislocations at the heterojunction interface. The SiGe and Ge epitaxial layers were grown in a cold-wall ultra-high-vacuum chemical-vapor-deposition system on Si. Immediately after growth each buffer layer was in situ annealed for 15 min at 750°C in order to further reduce the dislocation density. Then the

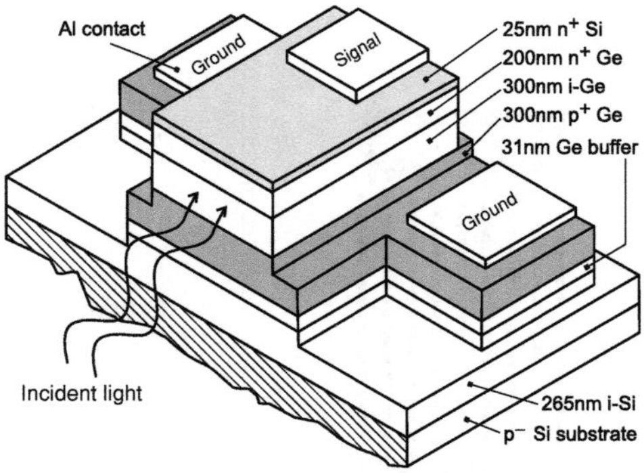

Fig. 3.12. Schematic device structure of Ge-on-Si PD [53]

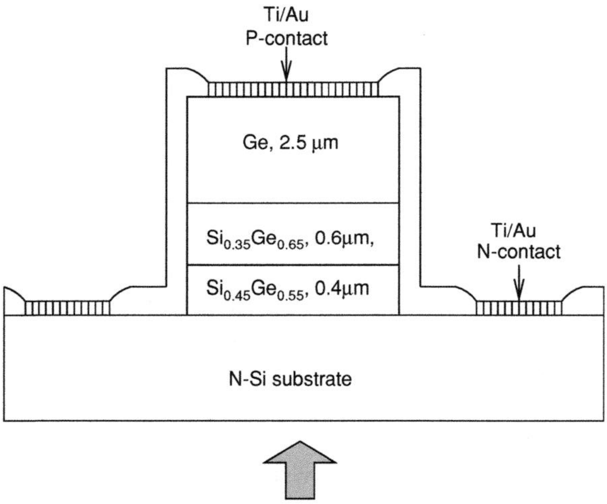

Fig. 3.13. Vertical Ge pin PD on two SiGe buffer layers [54]

temperature in the reactor was set to 400°C and the 2.5 μm thick Ge layer was grown. The measured sheet resistance of the grown films was 0.2 Ω/□. Mesas were formed by reactive ion etching and the mesa sidewalls were passivated with 200 nm SiO_2. Ti/Au contacts were deposited by electron beam evaporation and patterned by lift off. The total height of this PD above the Si substrate was 3.5 μm. The mesa diameter was 24 μm. To illuminate from the back side, the wafer was polished and 200 nm SiO_2 was deposited as an antireflection coating.

Bandwidths of 4.0 GHz at −3 V, 6.0 GHz at −5 V, 7.8 GHz at −7 V, and 8.1 GHz at −10 V were measured for 1.3 μm light. Dark currents of 0.06 and 1.07 μA were observed for the 24 μm mesa devices at reverse biases of 1 and 10 V, respectively. The responsivity was 0.57 A W^{-1} above 2 V reverse bias.

Metal–semiconductor–metal (MSM) photodetectors were also investigated, whereby the semiconductor material was Ge on Si [55]. Amorphous Ge thereby

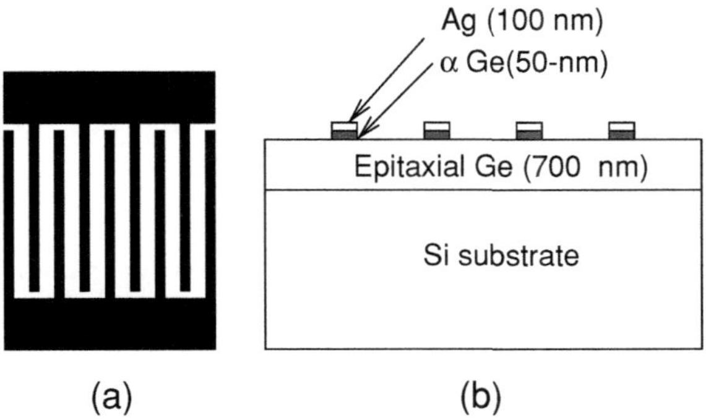

Fig. 3.14. MSM Ge PD on Si: (**a**) top view; (**b**) cross section [55]

was used to increase the barrier height of silver Schottky contacts on p-type Ge. Nevertheless, the dark current amounted to 7.5 μA at 3 V for a $25 \times 50\,\mu m^2$ active area. Figure 3.14 shows the cross section of this MSM photodetector.

The Si substrate was (100) oriented and with a resistivity of 5–25 Ω cm. Low-temperature epitaxial growth of a 700 nm thick crystalline Ge layer was done. This layer was p-type with an acceptor concentration of about $10^{17}\,cm^{-3}$. With a contact width of 1 μm and a spacing of 2 μm between these silver contact fingers a bandwidth of 4.3 GHz resulted for a reverse bias of 4 V. The external quantum efficiency was 14.3% ($0.15\,A\,W^{-1}$) without an antireflection coating compared to a higher responsivity of $0.24\,A\,W^{-1}$ at 1.3 μm [56].

In [57] these more efficient MSM Ge photodetectors also were described. They were grown at 600°C. A mobility of $1{,}200\,cm^2\,V^{-1}s^{-1}$ was reported for the Ge film which is only a factor of three lower than the value for bulk Ge. Due to the presence of defects in the Ge layer the carrier lifetime was estimated to be in the order of nanoseconds. The responsivity was $0.24\,A\,W^{-1}$ at 1.32 μm with a 1 V reverse bias. The internal quantum efficiency was reported to be 89%.

Other MSM photodetectors on 0.12–1.8 μm thick polycrystalline Ge-on-Si films deposited at temperatures from 25°C to 500°C were described in [57]. Such low-temperature Ge deposition was intended to enable postprocessing of silicon wafers and thereby integrability of Ge photodetectors with silicon electronics. Silver contacts were applied. Results on band alignment were reported. Figure 3.15 illustrates a valence band offset of 0.4 eV both for poly-Ge on n-type and p-type silicon.

As far as photosensitivity is concerned, devices based on amorphous Ge do not exhibit strong near-infrared (NIR) photoresponse above 1.2 μm. A carrier lifetime of 5 ns was reported and the responsivity at 1.32 μm was $16\,mA\,W^{-1}$. Due to the high conductivity of the poly-Ge corresponding to a narrow space-charge region the transport mechanism is mainly diffusion in the quasi-neutral zone with a rather short diffusion length of $L_D = 20$–30 nm.

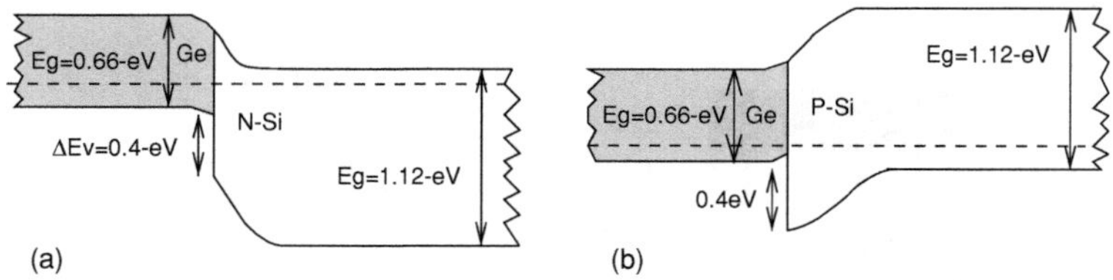

Fig. 3.15. Band alignment of poly-Ge on Si: (**a**) n-type Si; (**b**) p-type Si [57]

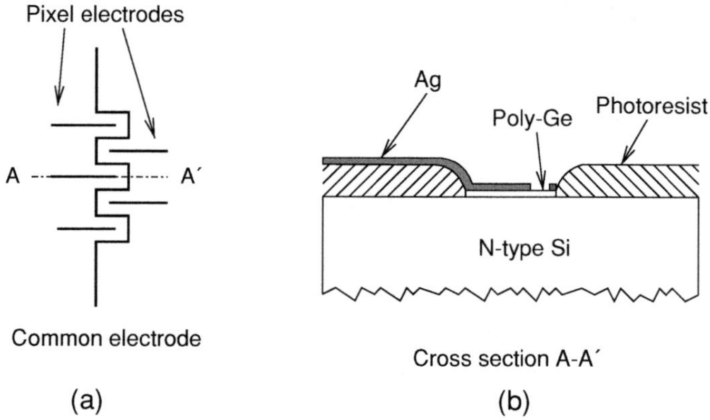

Fig. 3.16. Array of poly-Ge MSM PDs on Si: (**a**) top view; (**b**) cross section [57]

A linear array with 16 polycrystalline Ge photodetectors was demonstrated [57]. Figure 3.16 illustrates such an array with active pixel areas of $100 \times 100\,\mu m^2$. The responsivity of this detector array grown at 300°C was $16\,mA\,W^{-1}$ at $\lambda = 1.32\,\mu m$.

Meanwhile lateral poly-Ge detectors were monolithically integrated together with CMOS readout electronics [58]. Figure 3.17 depicts these advanced devices.

The detectors were fabricated by evaporation of Ge films at the end of a standard two-metal 0.7 µm CMOS process on substrates held at 300°C. An array of 64 detectors was realized. The Ge was evaporated in areas of 66 µm in square on n-wells used as cathode of the PD. All cathodes were connected. The p^+ diffusion areas embedded in the n-well provide ohmic contacts to the Ge anode layer and are electrically connected to the readout electronics. The resulting parasitic Si p^+–n junction, in parallel to the heterojunction PD, does not affect its behavior, owing to a negligible dark current and to lack of NIR sensitivity.

A dark current density of $3\,mA\,cm^{-2}$ at 1 V reverse bias was reported. This is about three orders of magnitude larger than for pure Si PDs. The maximum responsivity of the poly-Ge/Si PDs was $0.9\,mA\,W^{-1}$ at 1.3 µm for a reverse bias of 1 V. The authors mention that the responsivity can be improved by optimizing the Ge layer thickness (which was not given in [58]).

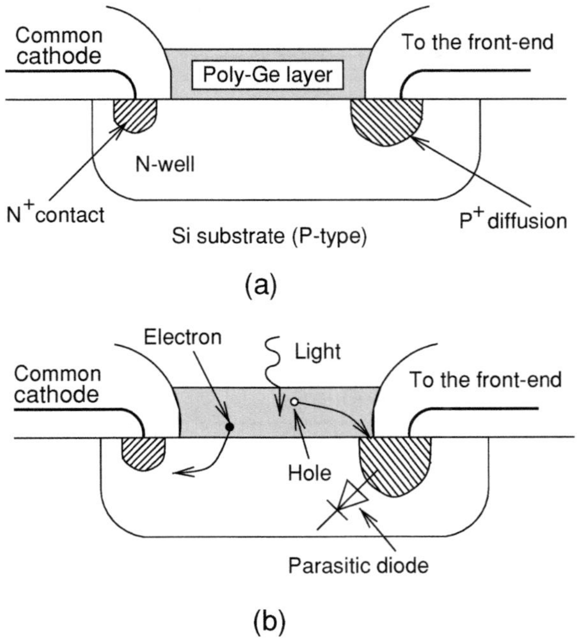

Fig. 3.17. (**a**) Schematic cross section of a monolithic poly-Ge on Si heterojunction photodetector; (**b**) description of carrier collection [58]

A back-illuminated voltage-tunable wavelength selective photodetector (VWP) on a 400 μm thick low-doped Si wafer (see Fig. 3.18) was suggested [57]. On a 500 nm thick SiGe (40% Ge) buffer layer, a 200 nm thick SiGe superlattice (SL, 145 periods of n-doped Si_6Ge_4, symmetrically strained) with a 2 nm Si N^+ cap layer and an Al contact was used. The equivalent circuit of the VWP contains two diodes, from which one is blocking and the other is conducting. Therefore, depending on the applied voltage the responsivity for different wavelengths (leading to different penetration depths of light) depends on the voltage.

After these Ge-on-bulk-Si PDs, Ge-on-SOI PDs will be described. SOI instead of a plane Si substrate was used to increase the Ge layer quality [59]. Figure 3.19 depicts such a lateral Ge photodetector reported similarly on bulk Ge [60].

The Ge layer was grown by ultra-high-vacuum chemical vapor deposition directly on an ultra-thin SOI layer (15 nm) after growth of a 30 nm Si buffer layer. At 350°C, first a 50 nm Ge seed layer was grown to suppress three-dimensional growth. Then a 400 nm thick layer was grown at 600°C. The density of threading dislocations then was reduced by thermal cyclic annealing by ramping ten times between 780 and 900°C for 6 min each. The threading dislocation density then was 10^8 cm^{-2}. The use of the ultra-thin Si layer limits the amount of Si and minimizes the diffusion of Si into the Ge layer.

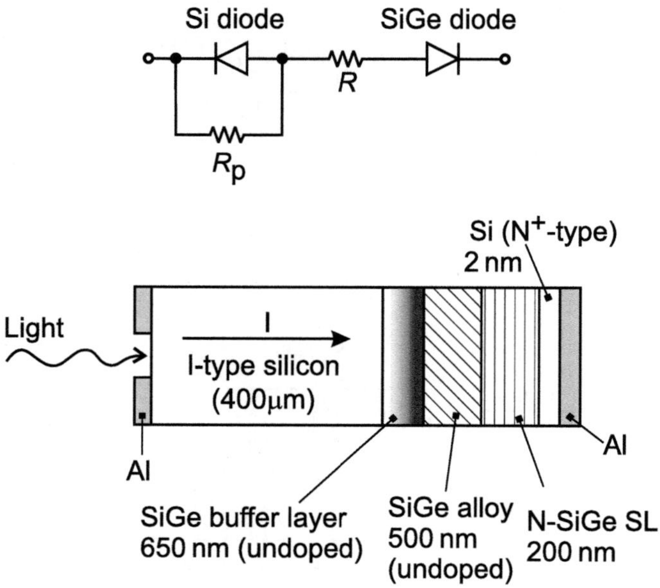

Fig. 3.18. Schematic of voltage-tunable wavelength selective photodetector (*bottom*) and electronic equivalent circuit (*top*) [57]

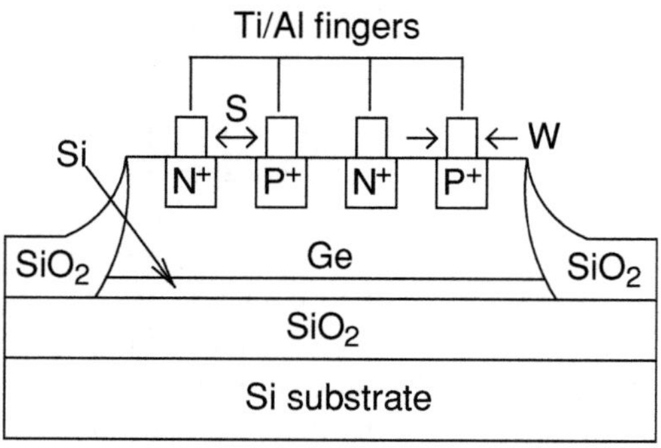

Fig. 3.19. Cross section of a lateral Ge-on-SOI pin PD [59]

The Ge layer was p-type with a mobility of 1,200 $cm^2 V^{-1} s^{-1}$ and with a carrier density of 10^{16} cm^{-3}. Mesas were etched and boron and As were implanted to form the p^+ and n^+ contacts, respectively. Ti/Al contacts were patterned onto the contact fingers. The electrode spacing S ranged from 0.3 to 1.3 μm, while the implant width W was 0.3 μm. No antireflection coating was used. Dark currents for 10×10 μm^2 devices were below 0.08 μA. The −3 dB bandwidths were 27, 23, and 19 GHz for spacings $S = 0.6$, 0.8, and 1.0 μm, respectively, at a reverse bias of 2 V. The detectors with $S = 0.6$ μm had an external quantum efficiency of 34% at 850 nm and 46% at 900 nm. Slightly higher quantum efficiencies of 38% and 52%, respectively, were reported in [61].

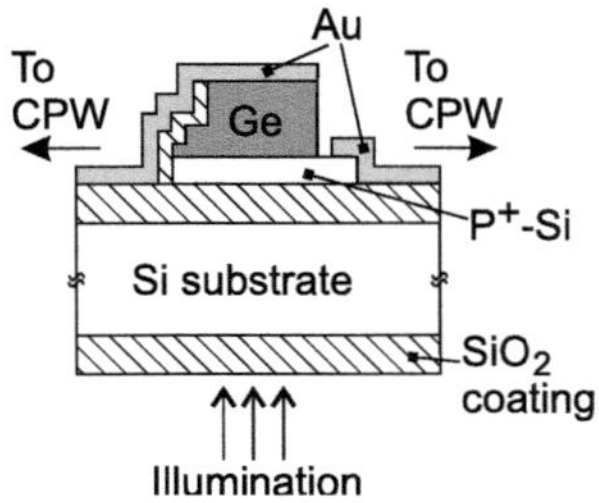

Fig. 3.20. Cross section of Ge-SOI Schottky PD [62]

An SOI structure was exploited in a resonant-cavity-enhanced (RCE) photodetector [62]. This back-illuminated detector is shown in Fig. 3.20.

A double-sided polished SOI substrate with an Si layer thickness of 340 nm and a buried oxide thickness of 200 nm was used. These layer thicknesses provide adequate back-illuminated reflectivity of 55% around the 1,550 nm wavelength region. Prior to Ge growth, boron implantation into the top Si layer was performed to obtain the bottom p-contact of the Ge photodetector. Then a 1,450-nm-thick Ge film was grown by a low-temperature Ge buffer layer technique. To reduce the threading dislocation density within the Ge layer, the Ge-SOI structure was cyclic annealed. Prior to the fabrication of the Ge photodetector, the Ge layer was etched back to a thickness resonant at 1,550 nm under back-illumination. Circular mesas were etched and Ti–Au metal contacts were then patterned to the oxide layer. On top of the mesas Au was used to form a Schottky contact and the mirror required to complete the RCE structure.

A reverse current of 380 nA was observed at 5 V for a 10 μm diameter Ge detector. A maximum transit-time limited 3 dB bandwidth of 12.8 GHz was reported for such a 10 μm diameter Ge detector at 1,550 nm and 4 V reverse bias. The RCE effect let to a quantum efficiency of 59% (0.73 A W^{-1}).

After these detectors SiGe waveguide detectors will be discussed in the following. Strained-layer superlattice GeSi/Si PDs for normal incidence near 1.3 μm were suggested [63]. The energy bands of this detector are shown in Fig. 3.21. The absorption region consists of 10 periods of GeSi and Si thicknesses of 10 and 40 nm, respectively. The bandgap of strained layers is considerably smaller than that of unstrained layers. Therefore strained-layer superlattices are interesting to increase the optical absorption coefficient at a small detector thickness.

The measured external quantum efficiency was 1% at 1.3 μm and 17% at 850 nm at a reverse bias of 4 V and with an antireflection coating. Therefore, such a detector is not very interesting for operation at 1.3 μm.

Another GeSi strained-layer superlattice detector was reported in [64]. Actually this detector was a waveguide detector. The light has to be coupled into this waveguide laterally. Figure 3.22 shows its device structure.

With a device length of 300 μm and lateral light incidence an internal quantum efficiency of 40% at 1.3 μm was reported. The modulation bandwidth exceeded 1 GHz. Leakage currents were below 3 μA [64].

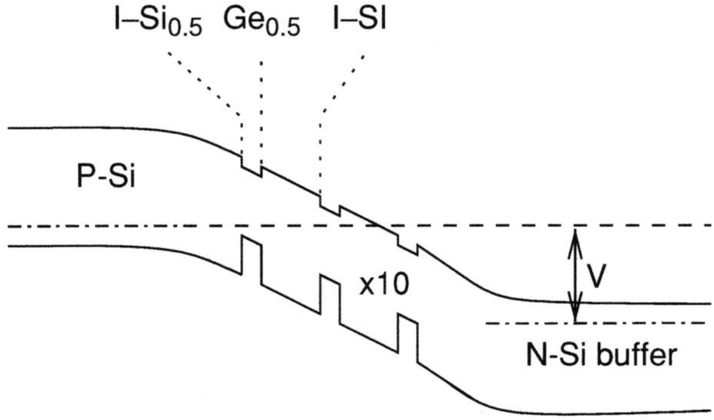

Fig. 3.21. Energy bands of GeSi superlattice PD [63]

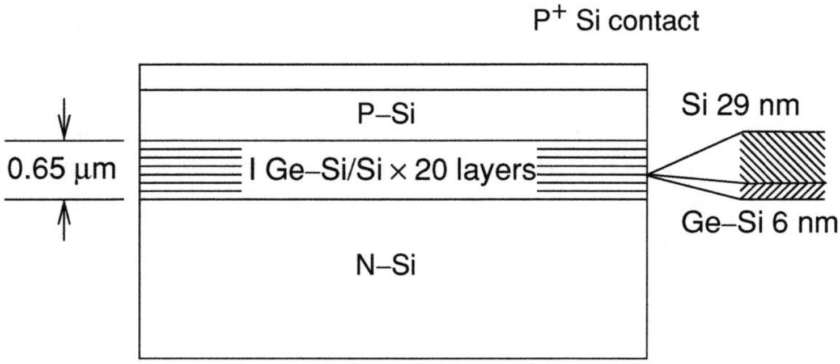

Fig. 3.22. Cross section of GeSi rib waveguide PD [64]

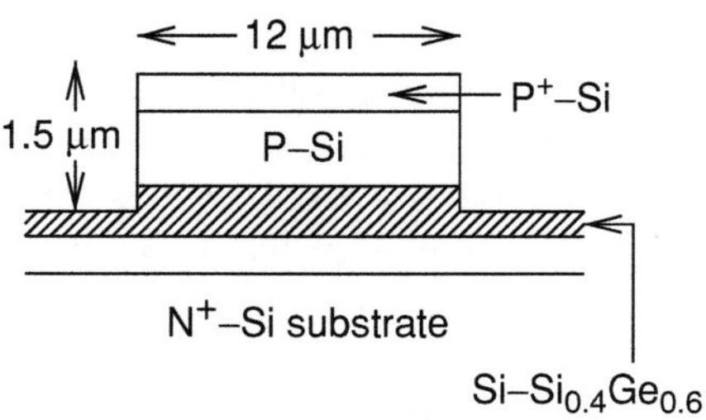

Fig. 3.23. Cross section of GeSi rib waveguide avalanche PD [65]

Avalanche GeSi rib waveguide photodetectors were investigated [65]. The cross section of such a detector is shown in Fig. 3.23.

A strained $Ge_{0.6}Si_{0.4}$ superlattice consisting of 20 layers was grown by molecular beam epitaxy on an n-type Si substrate. The length of the devices was 500 µm. The bandwidths was reported to exceed 8 GHz at an avalanche gain of 6. The breakdown voltage V_{br} was 32 V, where the avalanche gain achieved a factor of 12–17. The dark current, however, was with 20 µA quite large at a reverse bias of $0.8 \times V_{br}$.

Conclusion. There are promising investigations of Ge-on-Si photodetecotrs. Ge-on-Si detectors, however, are not yet available in commercial silicon technologies. Poly-Ge photodetectors being compatible with standard silicon technologies unfortunately possess a very low responsivity excluding them for BM applications. MSM photodetectors showing better responsivities at 1.3 μm are not yet available in standard silicon technologies. SOI is used for CMOS by IBM, however, to our knowledge it is not combined by Ge technology yet for commercial chip processing. Also RCE structures needing double-sided polished wafers or chips are not implemented in production silicon technologies yet. Integrated Ge detectors, therefore, are not available for the investigation of BM receivers.

3.4.4 SiGeC

The basic idea to overcome the limitations of SiGe alloys listed earlier in Sect. 3.4.1 is to add carbon (C) to the $Si_{1-x}Ge_x$ structures. The lattice constant of $Si_{1-y}C_y$ is smaller than that of Si offering the possibility to compensate for the larger lattice constant of $Si_{1-x}Ge_x$ and to reduce the mismatch to Si by the growth of $Si_{1-x-y}Ge_xC_y$ layers. Substitutional carbon levels up to 5% have been achieved by advanced growth techniques at temperatures of 400–650°C [66, 67], despite the very low solid solubility of less than 10^{-4} at all temperatures. The ability to adjust the strain in pseudomorphic layers because of the small size of the carbon atom, which compensates for the strain produced by 8–10 Ge atoms, has been verified [34]. $Si_{1-y}C_y$ layers on (100) Si have been shown to be under tensile strain, and the compressive strain for low C fractions in $Si_{1-x-y}Ge_xC_y$ on (100) Si has been reduced compared to $Si_{1-x}Ge_x$ [67]. Zero strain or even tensile strain has been observed for higher carbon fractions. Due to the reduction of strain as carbon is added in compressively strained $Si_{1-x-y}Ge_xC_y$ on (100) Si, the increase of the critical thickness as C is added has been demonstrated [68]. Let us conclude here that dislocation densities can be reduced due to the incorporation of small C fractions in $Si_{1-x-y}Ge_xC_y$ layers.

The bandgap of $Si_{1-y}C_y$ or $Si_{1-x-y}Ge_xC_y$ does not increase as fast with y as one might expect from the large bandgap of SiC or diamond [69, 70] due to large lattice distortions near C sites. Although the bandgap increases as C is added and the strain is reduced, the bandgap does not increase as fast as it would if the strain were reduced solely by reducing the Ge fraction [34]. An $Si_{1-x-y}Ge_xC_y$ layer, therefore, has a lower strain and a larger critical thickness than an $Si_{1-x}Ge_x$ layer with the same bandgap. For a layer with a desired bandgap of 100 meV less than that of Si, the critical thickness without C, i.e., of $Si_{0.86}Ge_{0.14}$, would be about 25 nm, for instance. With 1% carbon, i.e., $Si_{0.82}Ge_{0.17}C_{0.01}$, the critical thickness can be increased to about 70 nm.

For the application of $Si_{1-x-y}Ge_xC_y$ layers in devices, not only the bandgap itself but also the alignment of conduction and valence bands across the heterojunction interface is important. There is little conduction-band offset in $Si_{1-x}Ge_x$/Si interfaces if relaxed $Si_{1-x}Ge_x$ buffers are not used. The absence of a conduction-band offset [71] has been found for compressively strained $Si_{1-x-y}Ge_xC_y$ on (100) Si although there is no common agreement on this topic. Tensile-strained $Si_{1-y}C_y$ grown commensurate on a (100) Si substrate, however, possesses a smaller bandgap than Si and the conduction-band offset is increased allowing the fabrication of modulation-doped devices, which exploit two-dimensional electron gases with the electrons confined to the $Si_{1-y}C_y$ layer. $Si_{1-y}C_y$ infrared photodetectors, therefore, also seem feasible, although the effect of the C on the mobility of the $Si_{1-y}C_y$ layers is not yet known [34].

Many heterojunction and superlattice devices require the dopants to be placed with nanometer precision during the growth with respect to the heterojunctions, but the dopant atoms can diffuse during subsequent thermal processing for device fabrication. Especially during the integration of devices onto silicon integrated circuits, which is the final goal of $Si_{1-x}Ge_x$, $Si_{1-y}C_y$, and $Si_{1-x-y}Ge_xC_y$ research, some thermal processing is inevitable. Here the diffusion of the p-type dopant boron is the most critical problem. Small amounts of substitutional C fortunately can strongly reduce the extent of boron diffusion, especially when the boron diffusion results from Si self-interstitial injection during oxidation or implant annealing [72, 73]. Substitutional carbon is known to act as a sink for silicon self-interstitials, which are required for boron diffusion. The retarding effect of C on boron diffusion can be exploited already with carbon concentrations of $10^{19}\,cm^{-3}$. In a HBT with a boron-doped $Si_{0.795}Ge_{0.2}C_{0.005}$ base, for instance, the out-diffusion occurring in an $Si_{0.8}Ge_{0.2}$ base has been suppressed and the HBT characteristics has been considerably improved [73]. The reduction of boron diffusion in $Si_{1-x-y}Ge_xC_y$/Si superlattices also may allow the development of new optoelectronic devices. SiGeC photodetectors are still to be developed and are not available yet.

3.4.5 SiGe/Si pin Hetero Bipolar Transistor Integration

SiGe PDs were described above and in [74,75], however, the amplifiers in these references employed pure Si transistor devices. SiGe/Si technology has developed rapidly in recent years and it has been demonstrated that it outperforms Si technology in terms of the speed of transistors [76]. Therefore, it is advantageous to combine SiGe/Si photodetectors with SiGe/Si transistors [77].

Figure 3.24 shows the cross section of the first monolithically integrated SiGe/Si pin PD and heterojunction bipolar transistor (HBT) front-end photoreceiver. The pin-HBT structure was grown by one-step molecular beam epitaxy (MBE). Table 3.1 contains the layer compositions and doping concentrations of the SiGe-OEIC. The emitter and collector layers consist of Sb-doped Si. The base layer of the double heterojunction npn HBT structure is an $Si_{1-x}Ge_x$ alloy with a smaller bandgap than that of Si. The Ge mole fraction in the base layer is graded from $x = 0.1$ at the emitter side to $x = 0.4$

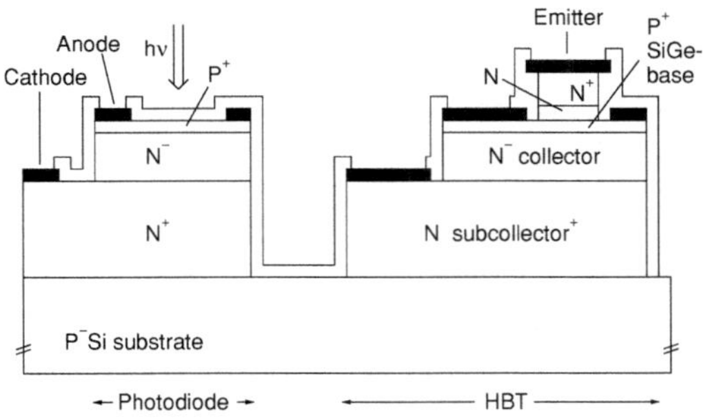

Fig. 3.24. Schematic cross section of an SiGe/Si pin HBT photoreceiver [77]

Table 3.1. Layer compositions and doping concentrations of an SiGe–Si pin-HBT OEIC

layer	material	type	doping (cm^{-3})	thickness (nm)
emitter contact	Si	n^+	1×10^{19}	200
emitter	Si	n	2×10^{18}	100
spacer	$Si_{0.9}Ge_{0.1}$	i		1
base	$Si_{1-x}Ge_x$ (x:0.1→0.4)	p^+	5×10^{19}	30
spacer	$Si_{0.6}Ge_{0.4}$	i		10
collector	Si	n^-	1×10^{16}	250
subcollector	Si	n^+	1×10^{19}	1500
substrate	Si	p^-	2×10^{12}	540 μm

at the collector side to accelerate the electrons traveling through the base towards the collector by a quasi-electric field. Spacer layers on both sides of the base minimize the effect of out-diffusion during epitaxy and processing.

The pin PD is formed by the Si n^+-subcollector, by the Si n^- collector, and the SiGe P^+ base layers of the HBT (see Fig. 3.24). The i-absorption layer is formed by the Si n^- collector due to the one-step MBE growth, which provides advantages over regrowth such as better planarity, simpler processing, and higher yield [77]. The MBE growth temperatures for the collector and emitter layers were 415°C. The base was grown at 550°C. The growth rate was only $0.2\,nm\,s^{-1}$ at these low temperatures.

The devices were isolated by mesa formation. The mesa size of the PD was $12 \times 13\,\mu m^2$. After MBE, the emitter contact was defined by evaporation, followed by emitter mesa formation with SF_6- and O_2-based dry and KOH-based wet etching. Minimal undercut and base over-etch were obtained by the two-step etch procedure, resulting in a low base access resistance. Then the base–collector mesa was formed by dry etching. The collector and pin cathode contacts were formed by evaporation on the exposed highly doped subcollector layer. Another etch step of the subcollector layer was applied for the separation of the devices. After this mesa isolation a PECVD SiO_2 layer was deposited and the pad contacts were opened.

The dark current of the PD was about 0.1 μA at $V_{pin} = 4$ V and 1 μA at $V_{pin} = 9$ V. For the pin PD with an i-layer thickness of 0.25 μm, a bandwidth of 450 MHz for $\lambda = 850$ nm and $V_{pin} = 9$ V was measured. The PECVD SiO_2 layer with a thickness of 1.1 μm served as an antireflection coating leading to a measured responsivity of 0.3 A W^{-1} and an external quantum efficiency of 43% for $V_{pin} = 5$ V. Here, the SiGe anode did not increase the quantum efficiency by a significant amount compared to an Si anode. For a large enhancement of the quantum efficiency a high Ge fraction in the intrinsic zone of the pin PD would have been necessary, which of course would introduce a high density of dislocations.

The bandwidth of 450 MHz was ascribed to the slow diffusion of carriers generated in the substrate and in the subcollector [77], because of the small i-layer thickness. We should not, however, accept this explanation, because the responsivity value of 0.3 A W^{-1} is too high for this explanation. It can be understood only when the cathode current, which is the sum of the subcollector/p^--substrate diode photocurrent and of the p^+-SiGe-anode/n-collector/n^+-subcollector diode photocurrent, was measured. The bandwidth then was not limited by carrier diffusion from the p^- substrate but by drift in the space-charge region of the subcollector/p^--substrate diode. The space-charge region extended far into the low doped p^--substrate with $N_A = 2 \times 10^{12}\,cm^{-3}$ and the measured bandwidth of 450 MHz for $V_{pin} = 9$ V seems possible. The reported increase in the responsivity and in the bandwidth with increased reverse bias also supports the new explanation with drift instead of carrier diffusion. This SiGe pin PD is not appropriate for 1.3 μm light due to the thin SiGe anode and the Si "intrinsic" zone.

The overall conclusion of this chapter is that SiGe- or Ge-photodetectors are not very appropriate or not yet feasible in SiGe BiCMOS technologies due to problems of metallization and planarization in the production. Poly-Ge detectors promise thinner Ge layers and less problems with lithography and metallization. Poly-Ge detectors, however, are not yet available in near standard Si processes. Integrated PDs in SiGeC technology also are not available yet.

4

Amplifiers

This short section gives an overview of the amplifiers in optical receiver front-ends. First the preamplifier basics are described and afterwards post amplifiers are discussed rudimentary. Due to the fact that the main noise source of an optical receiver is the preamplifier the description of the post amplifiers does not go into detail.

4.1 Preamplifier, Transimpedance Amplifier

The preamplifier is used to convert the incoming photocurrent into an output voltage, which is amplified by the following stages. The simplest way to do this conversion is a resistor between the PD output and the supply voltage as shown in Fig. 4.1.

The preamplifier is one of the determining parts concerning the sensitivity and bandwidth of an optical receiver. The sensitivity mainly depends on the responsivity of the PD and the input referred noise current of the circuit. Due to the fact that the output current of the PD is the smallest signal in the circuit, this point is the most sensitive concerning noise. The signal-to-noise ratio is most critical at the input node of the preamplifier. Therefore, the noise of the preamplifier is the dominating part of the input referred noise current. Again the noise of the resistor R and the first amplifying stage are the deciding factors.

The most interesting characteristics of the preamplifier, therefore, are the bandwidth and the input referred noise of the circuit. For the simple receiver shown in Fig. 4.1 the bandwidth is indirectly related to the capacitance of the input node and the resistor R. In this simple model, the capacitance of the input node consists of the capacitance of the PD, and the input capacitance of the main amplifier. To achieve a high bandwidth therefore the resistance R has to be small, as well as the capacitance of the input node. The noise of the circuit shown in Fig. 4.1 depends also on the resistor R, the capacitance

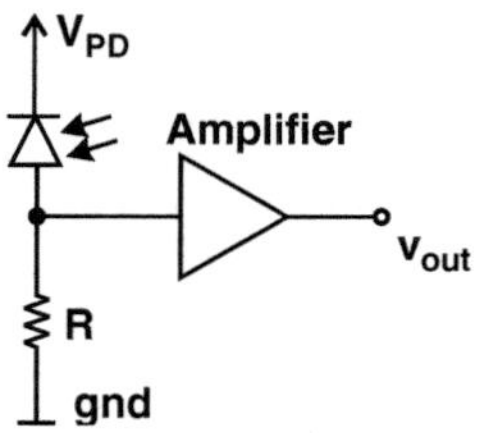

Fig. 4.1. Simplest possibility of the preamplifier

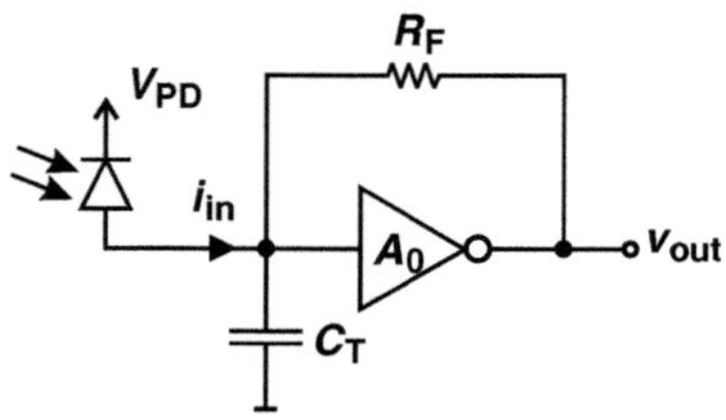

Fig. 4.2. Basic circuit of a TIA as preamplifier

of the input node and the first amplifier stage of the following amplifier. To achieve high bandwidths the resistor R must be small and therefore its noise current dominates the sensitivity of the optical receiver.

With a more complicated circuit, for example a TIA, a better performance can be achieved. Figure 4.2 shows a basic circuit of a TIA as a preamplifier.

The input referred noise current of the TIA also depends on the input node capacitance C_{T}, and the feedback resistor R_{F}. In the TIA circuit C_{T} consists of the capacitance of the PD and the input capacitance of the TIA, which nearly equals the input capacitance of the first amplifier stage. The advantage of a TIA compared to the simple circuit described before is the fact that the bandwidth is indirectly related to the resistor R_{F} divided by the open-loop gain A_{o} of the TIA $R_{\mathrm{F}}/A_{\mathrm{o}}$ (for more detail see Chap. 6). Therefore the noise can be decreased for a given bandwidth, because of a large resistor R_{F}.

The smaller C_{T} and the larger R_{F}, the higher is the sensitivity for a given responsivity (see Chap. 7). For a high bandwidth it is important to have small C_{T} and small R_{F}. Due to the fact that C_{T} is normally dominated by the PD capacitance, a tradeoff between sensitivity and bandwidth has to be found for the feedback resistor.

4.2 Main Amplifier, Limiting Amplifier

There are two different main amplifiers (MA): limiting and automatic gain control (AGC) amplifier. They have differential inputs and outputs to amplify

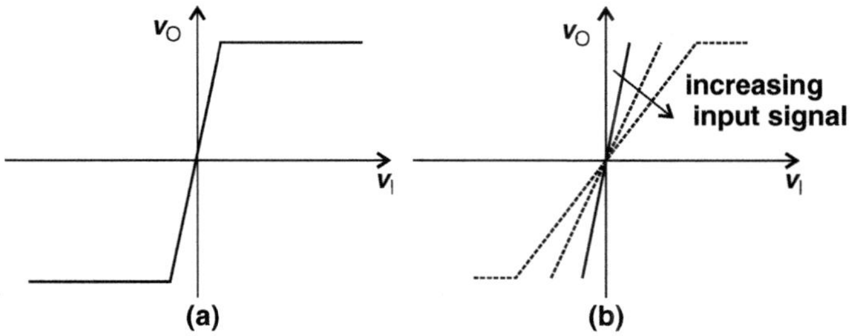

Fig. 4.3. DC transfer characteristics of (**a**) LA; (**b**) AGC amplifier [7]

a small input voltage signal from the TIA to an output signal which is sufficient for the reliable operation of the clock and data recovery.

4.2.1 Limiting Amplifier

Preamplifiers normally show a linear transfer function for input signals below a critical input signal. For input signals above the critical input signal there appear nonlinearities. The small frequency transfer function of the limiting amplifier (LA) is depicted in Fig. 4.3a. The region of linear operation and the area of dominating nonlinearities can be seen clearly. Due to the given power supply voltage the limiting effect of the LA appears naturally and no special design is necessary. Nevertheless, limiting effects like pulse-width distortion and delay variations must be minimized by a controlled limiting function.

Figure 4.4 shows an example of an LA designed by [78] in an 1 μm CMOS technology. It is a six-stage configuration in a fully differential topology which enables dc coupling between the single stages. The differential design has also the advantage of a good substrate-noise compression (see Fig. 4.4). The limiting amplifier stages themselves are designed in so-called Cherry–Hooper architecture which was introduced in bipolar technology by Cherry and Hooper [79] already in the 1960s. Figure 4.4b shows a CMOS version of this architecture. It is basically a series of transadmittance and transimpedance amplifiers, where the input voltage is first converted into an output current by the transadmittance amplifier and this current is again converted and amplified by the wideband TIA into an output voltage [78]. Several other designs were realized with different technologies, e.g., in SiGe HBT [80, 81], and in slight modifications, for example with the use of the shunt-peaking effect due to the inductances of the bond wires towards the supply voltage [82].

4.2.2 AGC Amplifier

Figure 4.3b shows the basic function of an automatic gain control amplifier (AGC). For large input signals the gain is reduced to keep the AGC in the linear region. This reduction is only possible in a defined region. Input signals above this region are limited too.

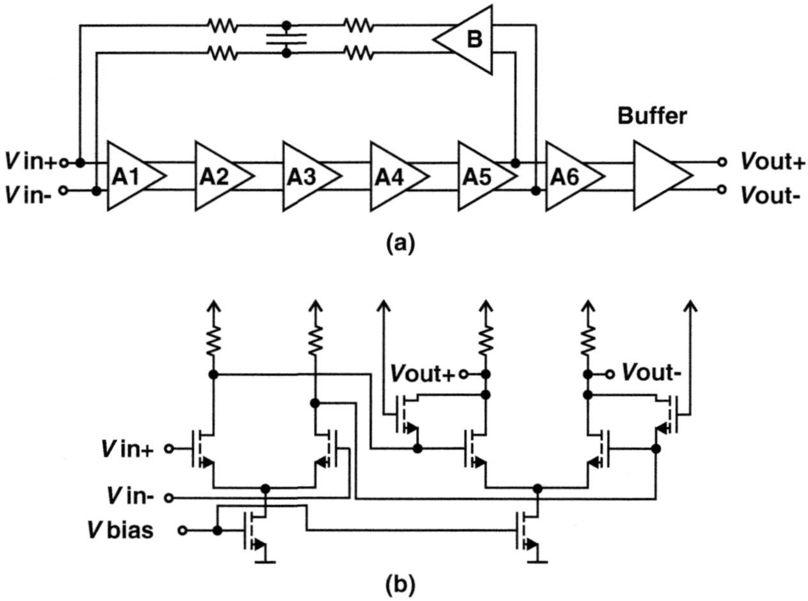

Fig. 4.4. Schematic of a limiting amplifier: (**a**) block diagram; (**b**) one amplifier stage [78]

Conventional preamplifiers, which control their transimpedance bit by bit, have problems receiving large signals with a low extinction ratio because of the nature of logarithmic amplifier operation [83–86]. In these preamplifiers, a current-bypass diode or current-bypass circuit is connected in parallel with the feedback resistor. When the output voltage of the transimpedance preamplifier exceeds the turn-on voltage of the bypass circuit, current flows through this circuit lowering the transimpedance. This type of transimpedance control was called bit-AGC (bit automatic gain control) in [8].

If a large input signal with a low extinction ratio is applied to such a bit-AGC preamplifier, the output waveform has a large dc part (see Fig. 4.5). The ac amplitude of the output signal is therefore reduced and it is difficult to decide between "0" and "1" properly. To solve the problem of output signal reduction, a novel circuit controlling the transimpedance cell by cell was proposed [8]. This type of control was called cell-AGC. The cell-AGC (see Fig. 4.6) is based on a bottom-level detector (BLD), a gain control circuit (GCC), a reset circuit, and an MOSFET connected in parallel with the feedback resistor.

BLD has quickly to detect the output signal of the three-stage TIA. A capacitor in BLD holds this bottom level. Depending on this level, the GCC generates a constant voltage during operation in a cell and determines the resistance of the MOSFET connected in parallel to R_{F} via its gate source voltage V_{GS}.

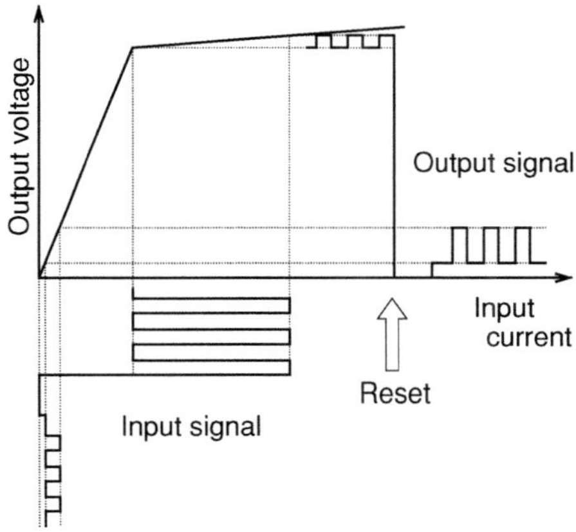

Fig. 4.5. Operation of bit-AGC [8]

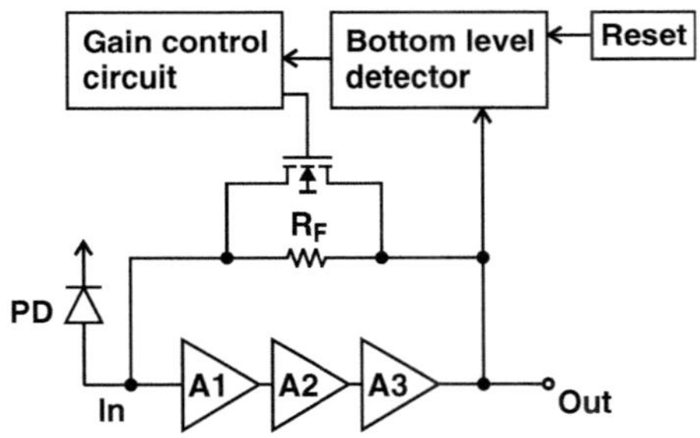

Fig. 4.6. Block diagram of three-stage BM receiver [8]

Between two cells, the reset signal is applied to BLD, the hold capacitor is charged, and the output of GCC returns to its initial state [8]. To increase transmission quality, a fast response from BLD and GCC is necessary during the first bit in the new cell. The response of cell-AGC to input signals with different power levels is illustrated in Fig. 4.7. For a large-signal cell, the transimpedance is set to G1 instead of to G2 for a weak-signal cell. Because the output voltage is proportional to the input current, the dc background level for the "0" level is not as large as that in bit-AGC. With the help of this circuit, therefore, a decision between "0" and "1" can be performed properly and burst cells with a low extinction ratio can be received.

A 0.25 μm CMOS technology was selected for the realization of the cell-AGC BM receiver aiming at noise reduction for a high sensitivity. A feedback resistor with a value of 40 kΩ was implemented in the transimpedance preamplifier. The transconductance g_{m} of the input MOSFET was set to about 50 mS. Stability is the major problem to be solved for three-stage TIA. Therefore, the influence of changes in g_m with process deviations on the stability has to be suppressed and the value of g_{m} has to be kept close to the minimum for a low power consumption. This requirement can be met with the series

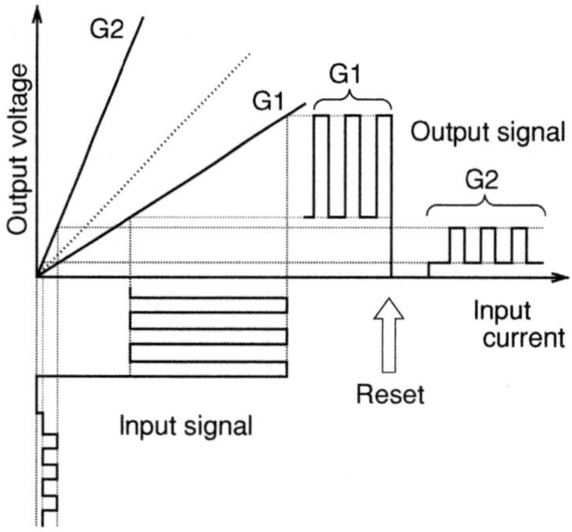

Fig. 4.7. Operation of cell-AGC [8]

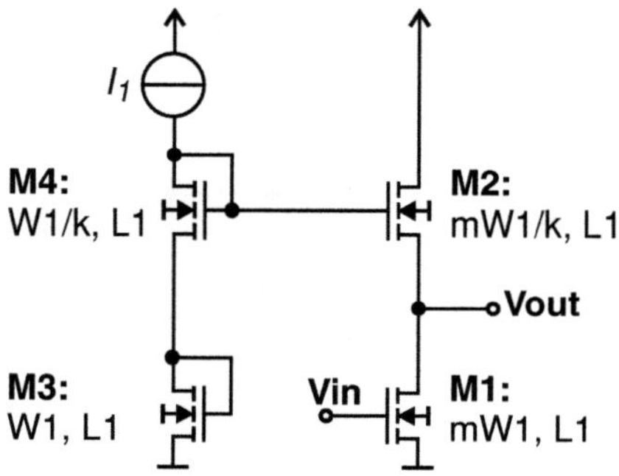

Fig. 4.8. Configuration of g_{m}-over-g_{m} amplifier stage [8]

MOSFET load configuration shown in Fig. 4.8. Each of the three amplifier stages in Fig. 4.6 consists of the circuit shown in Fig. 4.8.

Due to the condition of $V_{\mathrm{in}} = V_{\mathrm{out}}$ for negative feedback, V_{GS} of M1 is equal to V_{GS} of M3 and V_{GS} of M2 is equal to V_{GS} of M4. The transconductances are given by:

$$g_{\mathrm{m1}} = \sqrt{2\mu_{\mathrm{n}} C_{\mathrm{ox}} \frac{mW_1}{L_1}, I_1} \quad g_{\mathrm{m2}} = \sqrt{2\mu_{\mathrm{n}} C_{\mathrm{ox}} \frac{mW_1}{kL_1} I_1}, \tag{4.1}$$

where μ_{n} is the electron mobility, C_{ox} is the gate oxide capacitance per area, I_1 is a reference current, and k is the ratio of the gate widths. The gain of this amplifier is equal to the ratio of the transconductances. This amplifier therefore is called g_{m}-over-g_{m} amplifier.

$$G = \frac{g_{\mathrm{m1}}}{g_{\mathrm{m2}}} = \sqrt{k}. \tag{4.2}$$

The gain G of such an amplifier stage only depends on the parameter k. It is completely independent from process deviations and operating temperature. The current flowing through M1 is $m \times I_1$, such that the gain G and g_m can be stabilized also under power supply deviations. As G is stabilized, the bandwidth $f_{3\,\mathrm{dB}}$ is kept stable, i.e., constant, due to the relation

$$f_{3\,\mathrm{dB}} = \frac{1+G}{2\pi R_F C_F}.$$

The bandwidth – important for constant integrated noise – therefore can be designed to be stable around the optimum bandwidth.

The preamplifier in a 0.25 µm CMOS technology with an chip size of $1.3 \times 1.12\,\mathrm{mm}^2$ was characterized at a supply voltage of 2.5 V ± 5% with an external PD having a responsivity of $0.9\,\mathrm{A\,W}^{-1}$ ($\lambda = 1.3\,\mu\mathrm{m}$) and a capacitance of 0.6 pF [8]. The bandwidth varied from 122 to 164 MHz in the operating temperature range from −40 to +85°C and also due to remaining R_F and C_F deviations. The power consumption was 60–71.4 mW at 2.5 V supply voltage. A sensitivity of −39.3 dBm for a BER of 10^{-10} at a data rate of $156\,\mathrm{Mb\,s}^{-1}$ was reported [8]. The maximum optical input power was −6 dBm.

5

Integrated Circuit Technology

In the following chapter the fundamentals of SiGe heterojunction bipolar transistors and deep-sub-micron CMOS technologies will be discussed. Bipolar processes and standard CMOS processes are discussed in [13, 87] and therefore will not be discussed again.

5.1 SiGe Heterojunction Bipolar (HBT)

The first bandgap-engineered device realized in Si is the SiGe HBT. It combines the transistor performance comparable to III–V technologies and the process maturity, integration depth, the rate of yield as well as the cost effectiveness of standard Si processes [88].

Figure 5.1 shows a schematic cross section of an SiGe HBT. Deep-trench isolation and multilevel metallization is not depicted in Fig. 5.1 to get a clear view. The structure is self-aligned and planar with conventional polysilicon emitter contact and silicided extrinsic base [88]. The isolation is done by shallow- and deep-trench isolation.

Due to the compatibility with Si CMOS fabrication, with its inherently significant thermal cycle, the resulting doping profile of the emitter, the base and the collector of a HBT looks more similar to that of a conventional ion-implanted base Si bipolar transistor than to that of a conventional III–V-material HBT [88]. This means the base in the SiGe HBT of an SiGe BiCMOS technology is not as heavily p-type doped as in a III–V HBT [88].

The content of Ge in the base area reduces the bandgap energy of the material. The bandgap energy is decreasing with increasing Ge concentration. This fact can be used to build a decreasing bandgap in the direction of the electron flow (see Fig. 5.2a) in case of an increasing Ge concentration towards the collector [89] like depicted in Fig. 5.2b.

The electrons are injected in the emitter of the transistor. The bandgap is reduced, compared to an Si bipolar transistor, due to the Ge concentration Ge_E on the emitter-end of the base. The electric field – because of the increasing concentration of Ge – accelerates the electrons further towards the

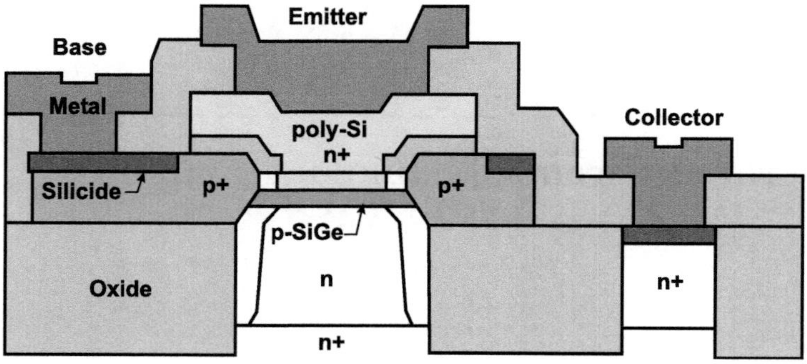

Fig. 5.1. Schematic cross section of an SiGe heterojunction bipolar transistor (HBT) [88]

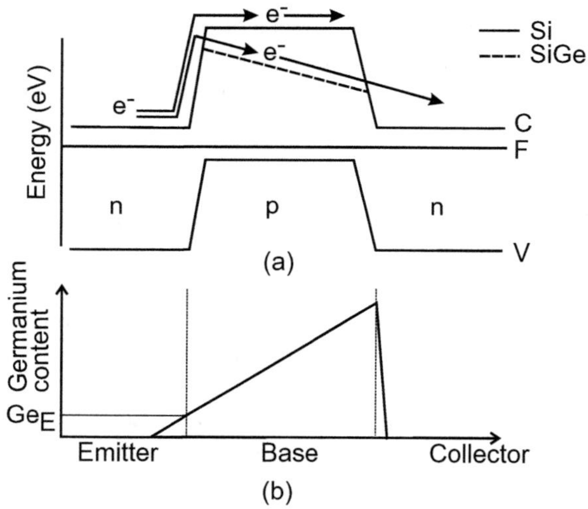

Fig. 5.2. (**a**) Conduction band and valence band energy levels for Si bipolar and SiGe HBT; (**b**) graded Ge content of IBM SiGe HBT [89]

collector. This effect speeds up the transport of the electrons from emitter to collector and enables therefore higher-frequency operation [89]. The current gain β is also increased by the introduction of Ge into the base [88]. Furthermore, the early voltage is increased by the graded Ge concentration, because the intrinsic carrier concentration is modulated across the base and therefore the output characteristic of the HBT is improved.

For IBM technologies the HBT cut-off frequency could be increased from 47 GHz in the 0.5 µm generation to 210 GHz in the 0.13 µm generation [89]. The 0.18 µm IBM SiGe BiCMOS process shows a cut-off frequency of 100 GHz for the bipolar transistors [90]. The minimum noise figure is below 0.6 dB in the frequency range between 3 and 10 GHz and the $1/f$ corner frequencies are below 400 Hz.

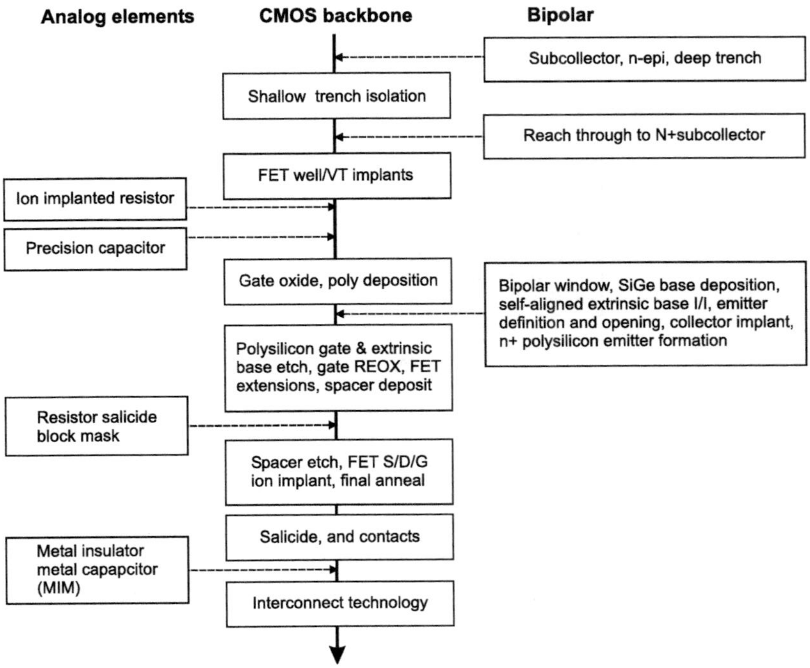

Fig. 5.3. Schematic diagram of process flow of 0.5 μm SiGe BiCMOS process in base-during-gate production [91]

There are two types of process flows for the production of SiGe HBTs. Both use a standard CMOS process flow and add several process steps. For 0.5 μm technologies, for example, IBM used the approach to share thermal cycles and layers for the SiGe bipolar integration. This approach reduces the complexity of the process. This leads to so-called base-during-gate (BDGate) production [91] (see Fig. 5.3).

With the further development of CMOS processes, another approach is pursued. The bipolar elements are formed after some CMOS elements without sharing layers or thermal cycles. This approach is called base-after-gate production [91]. Due to the fact that analog parts are already integrated in complex, modern deep-sub-micron processes, no more additional analog elements are necessary. Figure 5.4 shows the process flow of the base-after-gate (BAGate) production like it is done in 0.18 and 0.13 μm technologies.

5.2 Deep-Sub-Micron CMOS

Deep-sub-micron CMOS processes were used for our own designs of this book over the years. One design was done in 0.18 μm CMOS technology, the others were designed in 0.12 μm CMOS process. This section mentions the principal facts of the deep-sub-micron CMOS process.

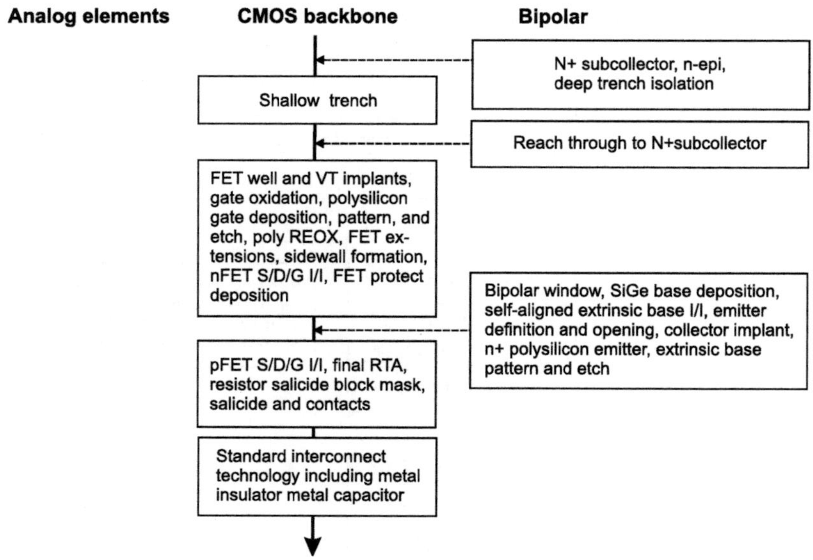

Fig. 5.4. Schematic diagram of process flow of 0.13 μm SiGe BiCMOS process in base-after-gate production [91]

The 120 nm CMOS process is often also the called 130 nm CMOS process although the minimum gate length is 120 nm, because the milestone in the IRTS roadmap [92] is defined at 130 nm gate length and therefore the allocation is done using the name 130 nm CMOS process. In this book, the process is called 120 nm CMOS process according to the actual minimum gate length.

These technologies are standard in digital designs. Therefore the process tolerances are rather high, which does not matter for digital designs. In system-on-chip solutions the analog and the digital part of the system can be integrated in the same technology, in this example an optical receiver is designed in the digital CMOS process and can be integrated with the digital signal processing part. Although deep-sub-micron CMOS processes offer analog extensions, they are not used in the designs presented in Chap. 9 to minimize the process costs. This means, on the other hand, that the design has to deal with the nonideal properties of the devices.

Figure 5.5 displays a cross section of a die structure of a deep-sub-micron CMOS process. The example 130 nm technology [93], again the actual drawn gate length is 120 nm, offers six copper layers for wiring. The upper two copper layers are thicker and therefore they can handle more current.

The process provides enhancement mode n-channel and p-channel MOS transistors. The used processes are twin-well processes on nonepi p-substrate. Therefore, n-MOSFETs are directly implemented in the substrate surrounded by a p-implant area, while the p-MOSFET is situated in an n-well. This leads to the back-gate effect for n-MOSFETs, where the source is not connected to VSS, e.g., in a differential amplifier. The n-well potential, and therefore the

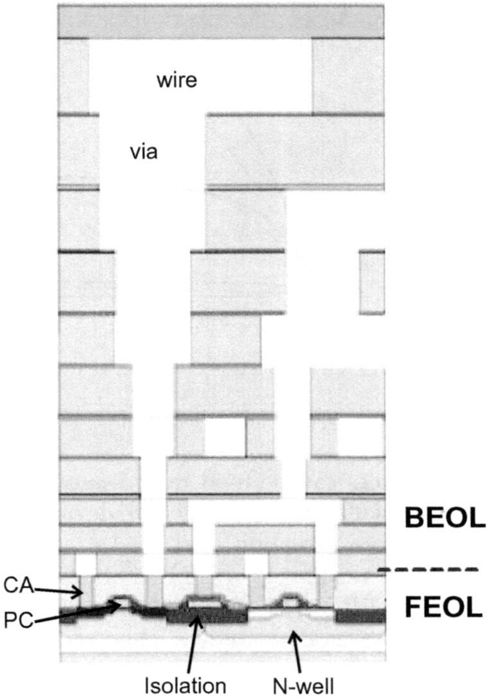

Fig. 5.5. Cross section of a die structure (130 nm) [93]

bulk of the p-MOSFETs, is user defined and therefore the back-gate effect can be prevented when connecting bulk and source together.

The circuits are designed with regular threshold (V_t) devices, which are the standard devices in this process. It also offers high-V_t and low-V_t transistors with the same physical oxide thickness and therefore they are useable for the same supply voltage of 1.5 V. I/O transistors with an operation voltage of up to 2.5 V and analog transistors with restricted use up to 3.3 V are also provided with an effective oxide thickness of more than two times the thickness of the oxide of the standard transistors. These transistors can be used to fulfill special demands on external supply voltages and output resistances.

The low-V_t transistors show the lowest threshold voltage and the lowest output resistance of the three types of transistors with standard effective oxide thickness. They can be used as analog devices to drive higher drain currents at the same VDD compared to the regular-V_t devices and therefore have a higher g_m. The disadvantages are the low output resistance and the high subthreshold current.

High-V_t transistors have a high threshold voltage and also a high output resistance. They are used for digital designs with low power consumption due to the high output resistance and low subthreshold current.

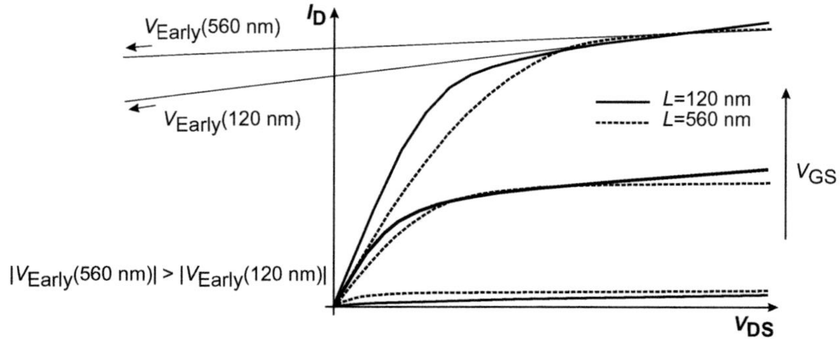

Fig. 5.6. Principal transistor characteristics for 560 nm gate length and 120 nm gate length [94]

High-V_t or low-V_t devices can be used in the same chip. For example the high-V_t transistor device is a regular-V_t FET-like device with an additional V_t adjust implant and therefore two additional masks.

In Fig. 5.6 two typical regular V_t n-FET characteristics are shown. The dashed curves show an n-FET with $L = 560\,\text{nm}$, which is the longest gate length, for which transistor models are available from the rf-CMOS library. Thin lines extend the slope of the curve in the saturation region towards negative numbers of drain-source voltage. An n-FET with $L = 120\,\text{nm}$ is depicted as solid curves (both characteristics are normalized to approximately the same maximum drain current). The achieved values of the early voltage, which are not depicted in the figure, due to the fact that the scaling would impede seeing any details in the actual transistor characteristics, are small for short gate lengths. It can be clearly seen that the early voltage decreases for 120 nm gate length compared to 560 nm gate length. From this figure it is clear that short FETs are particularly bad for analog circuits due to their low Early voltage. The usual assumption of a drain current I_D approximately independent of V_{DS} in the saturation region is wrong which results in a very low small-signal output resistance r_{DS}, as shown in the simplified equation (5.1)

$$r_{DS} = \frac{V_{DS} + V_{Early}}{I_D}. \tag{5.1}$$

Further devices, next to the FETs, are ohmic resistors and capacitances. The resistors are sheet resistors of different materials with different Ohm per square, e.g., polysilicon resistors for high Ohm per square, or diffusion-resistors for resistors with small values with low Ohm per square. Capacitances usually are implemented as parallel metal planes, or stacks of them. Very high capacitances per area unit μm^2 can be achieved with so-called *n-FET in n-well capacitors*, where the gate dielectric is used for the capacitance. One plate of the capacitor is formed by the n-well and the other one is formed by the gate of the n-MOSFET, which is placed in the n-well. The gate shows a high length and a width in the same order of magnitude, to achieve a very large area. These capacitances have the disadvantage that the value of the capacitance depends on the potential of the plates, but this disadvantage does not account for supply filtering. Here, it is important to achieve a capacitance value per area unit μm^2 as high as possible.

Large process variations in this process are no problem for digital circuits. For analog circuits these variations have a bad influence. The saturation current of a FET can vary up to 24% for different process variations at constant V_{GS} and V_{DS}. Resistor values vary by ±10% to ±15% and capacitors vary by ±13%. This complicates the design of the chip.

For high-frequency applications a library with specialized components is provided. This *rfcmos* library contains n-MOS and p-MOS transistors with a well-defined layout. The FETs have a predefined layout with special length and width. The transistor itself is the same regular-V_t transistor as known from the common library, but the layout is implemented in the simulation model of the FET. This library offers the possibility to do accurate simulations of the circuits up to several GHz.

6

Transimpedance Amplifier Theory

This chapter deals with the basic calculations for a transimpedance amplifier (TIA). First, theoretical considerations of feedback theory are made in Sect. 6.1. Afterwards ideal TIAs are discussed in Sect. 6.2 as well as nonideal TIAs in the following sections. Finally the stability of TIAs is discussed.

6.1 Feedback Theory

In this section the shunt–shunt feedback as well as input and output resistance are discussed theoretically on the base of an abstract network.

6.1.1 Shunt–Shunt Feedback

When the feedback in a transimpedance amplifier only consists of an ohmic resistor, the name transresistance amplifier can be used instead of transimpedance amplifier. Such a transresistance amplifier is formed by a shunt–shunt feedback configuration (Fig. 6.1). The parameters i_{S} and R_{D} are the photocurrent and the parallel resistance of a photodiode. The input port voltages are the same and the output port voltages are the same for the amplifier and feedback two-ports. The y-parameters, therefore, are appropriate for analyzing this shunt–shunt feedback configuration [95].

The amplifier and the feedback network are represented by their individual y-parameters:

$$\begin{aligned} i_1^{\mathrm{A}} &= y_{11}^{\mathrm{A}} v_1 + y_{12}^{\mathrm{A}} v_2, \\ i_2^{\mathrm{A}} &= y_{21}^{\mathrm{A}} v_1 + y_{22}^{\mathrm{A}} v_2, \end{aligned} \tag{6.1}$$

and

$$\begin{aligned} i_1^{\mathrm{F}} &= y_{11}^{\mathrm{F}} v_1 + y_{12}^{\mathrm{F}} v_2, \\ i_2^{\mathrm{F}} &= y_{21}^{\mathrm{F}} v_1 + y_{22}^{\mathrm{F}}, v_2, \end{aligned} \tag{6.2}$$

where the superscript A indicates the amplifier and the superscript F indicates the feedback network.

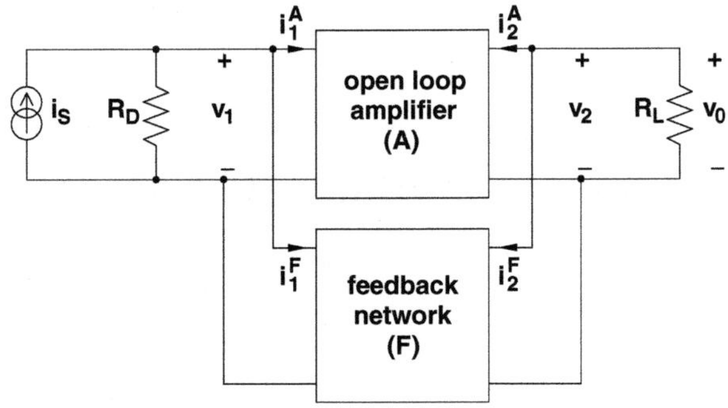

Fig. 6.1. Shunt–shunt feedback amplifier [95]

The total input current i_1 and the total output current i_2 can be written as

$$\begin{aligned} i_1 &= i_1^{\mathrm{A}} + i_1^{\mathrm{F}}, \\ i_2 &= i_2^{\mathrm{A}} + i_2^{\mathrm{F}}. \end{aligned} \tag{6.3}$$

Substituting (6.1) and (6.2) into (6.3) yields the two-port description for the shunt–shunt feedback amplifier:

$$\begin{aligned} i_1 &= (y_{11}^{\mathrm{A}} + y_{11}^{\mathrm{F}})v_1 + (y_{12}^{\mathrm{A}} + y_{12}^{\mathrm{F}})v_2, \\ i_2 &= (y_{21}^{\mathrm{A}} + y_{21}^{\mathrm{F}})v_1 + (y_{22}^{\mathrm{A}} + y_{22}^{\mathrm{F}})v_2. \end{aligned} \tag{6.4}$$

A more compact notation can be achieved by defining

$$y_{\mathrm{ij}}^{\mathrm{T}} = y_{\mathrm{ij}}^{\mathrm{A}} + y_{\mathrm{ij}}^{\mathrm{F}}$$

and

$$\begin{aligned} i_1 &= y_{11}^{\mathrm{T}}v_1 + y_{12}^{\mathrm{T}}v_2, \\ i_2 &= y_{21}^{\mathrm{T}}v_1 + y_{22}^{\mathrm{T}}v_2, \end{aligned} \tag{6.5}$$

because the corresponding parameters of both networks again appear together in (6.4). Since the input current is determined stronger by the feedback network than by the amplifier output and since the amplifier (and not the feedback network) drives the load, it is allowed to assume

$$\begin{aligned} y_{12}^{\mathrm{F}} &\gg y_{12}^{\mathrm{A}}, \\ y_{21}^{\mathrm{A}} &\gg y_{21}^{\mathrm{F}}, \end{aligned} \tag{6.6}$$

and (6.5) can be simplified to

$$\begin{aligned} i_1 &= y_{11}^{\mathrm{T}}v_1 + y_{12}^{\mathrm{F}}v_2, \\ i_2 &= y_{21}^{\mathrm{A}}v_1 + y_{22}^{\mathrm{T}}v_2. \end{aligned} \tag{6.7}$$

The closed-loop gain of the shunt–shunt feedback amplifier considering the effects of R_{D} and R_{L} can now be found with the help of (6.7). At the input port and output port in Fig. 6.1, v_1 and i_1 plus v_2 and i_2 are related by

$$\begin{aligned} i_1 &= i_{\mathrm{S}} - v_1 G_{\mathrm{D}}, \\ i_2 &= -G_{\mathrm{L}} v_2 \,. \end{aligned} \tag{6.8}$$

Substituting (6.8) into (6.7) yields

$$\begin{aligned} i_{\mathrm{S}} &= (G_{\mathrm{D}} + y_{11}^{\mathrm{T}})v_1 + y_{12}^{\mathrm{F}} v_2, \\ 0 &= y_{21}^{\mathrm{A}} v_1 + (y_{22}^{\mathrm{T}} + G_{\mathrm{L}})v_2 \,. \end{aligned} \tag{6.9}$$

The closed-loop transresistance can be obtained from (6.9) by solving for v_2 in terms of i_{S}:

$$A_{\mathrm{TR}} = \frac{v_2}{i_{\mathrm{S}}} = \frac{y_{21}^{A}}{y_{21}^{\mathrm{A}} y_{12}^{\mathrm{F}} - (G_{\mathrm{D}} + y_{11}^{\mathrm{T}})(y_{22}^{\mathrm{T}} + G_{\mathrm{L}})} \,. \tag{6.10}$$

Rearranging (6.10) into the standard form for a feedback amplifier gives

$$A_{\mathrm{TR}} = \frac{v_2}{i_{\mathrm{S}}} = \frac{\dfrac{-y_{21}^{\mathrm{A}}}{(G_{\mathrm{D}} + y_{11}^{\mathrm{T}})(y_{22}^{\mathrm{T}} + G_{\mathrm{L}})}}{1 + \dfrac{-y_{21}^{\mathrm{A}}}{(G_{\mathrm{D}} + y_{11}^{\mathrm{T}})(y_{22}^{\mathrm{T}} + G_{\mathrm{L}})} y_{12}^{\mathrm{F}}} = \frac{A}{1 + A\beta_{\mathrm{F}}}, \tag{6.11}$$

where

$$A = \frac{v_{\mathrm{O}}}{i_{\mathrm{S}}} = -\frac{y_{21}^{\mathrm{A}}}{(G_{\mathrm{D}} + y_{11}^{\mathrm{T}})(y_{22}^{\mathrm{T}} + G_{\mathrm{L}})} \quad \text{and} \quad \beta_{\mathrm{F}} = y_{12}^{\mathrm{F}}, \tag{6.12}$$

where the feedback parameter β_{F} has the dimension Ω^{-1} [95]. These two equations are interpreted in Figs. 6.2 and 6.3 for the case of shunt–shunt feedback. Figure 6.2 shows the feedback amplifier with an explicit representation

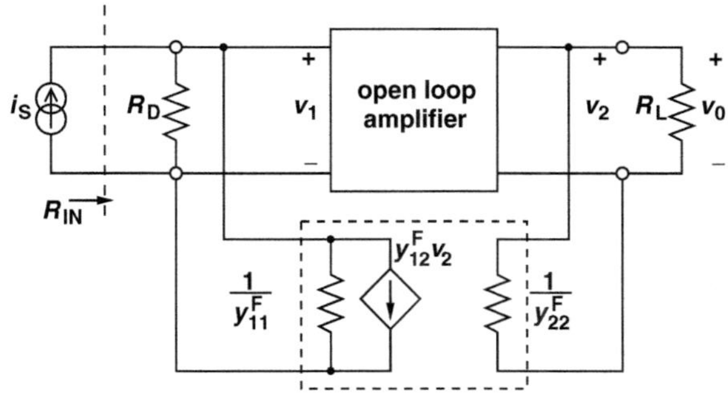

Fig. 6.2. Illustration of the amplifier described by (6.11) and (6.12)

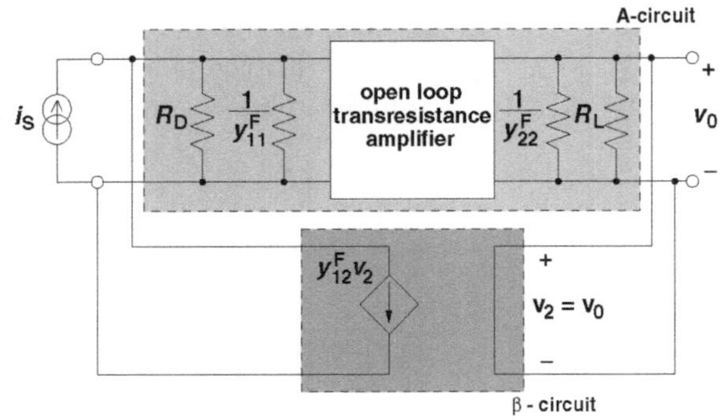

Fig. 6.3. Illustration of the feedback circuit described by (6.11) and (6.12) [95]

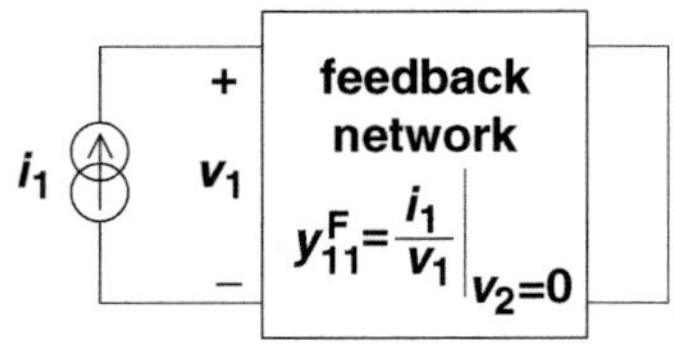

Fig. 6.4. First feedback circuit

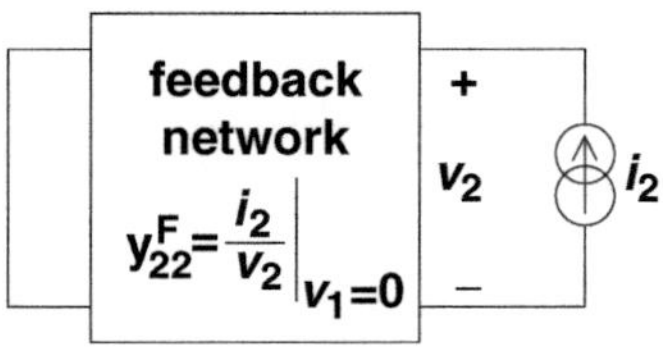

Fig. 6.5. Second feedback circuit

of the two-port parameters of the feedback network with $y^F_{21} = 0$. According to (6.11) and (6.12) the gain of the amplifier A has to be calculated including the effects of y^F_{11}, y^F_{22}, R_D ($R_D = 1/G_D$), and R_L ($R_L = 1 / G_L$).

A schematic representation of these equations is given by redrawing the amplifier as in the circuit in Fig. 6.3. The position of the feedback circuit elements y^F_{11} and y^F_{22} has been changed, but the overall circuit is once again the same. The amplifier A-circuit now includes y^F_{11}, y^F_{22}, R_D, and R_L, whereas the feedback network consists only of y^F_{12}.

Figures 6.4–6.7 illustrate the analysis technique in more detail. The three required y-parameters of the feedback network are found based on their individual definitions [95].

The determination of the y-parameter y^F_{11} with a shorted output is illustrated with Fig. 6.4. Figure 6.5 depicts the definition of the y-parameter y^F_{22}, where the input is shorted. Figure 6.6 shows the definition of the y-parameter y^F_{12}.

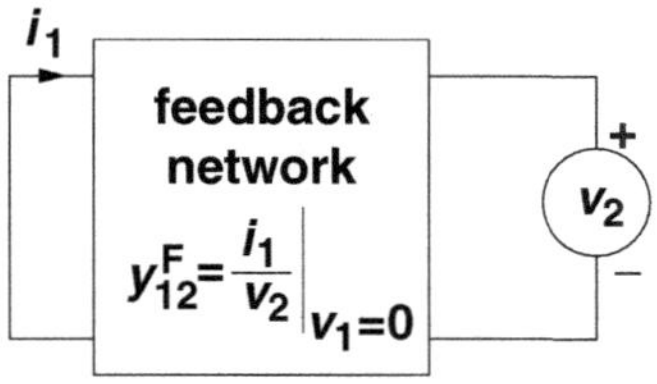

Fig. 6.6. Third feedback circuit

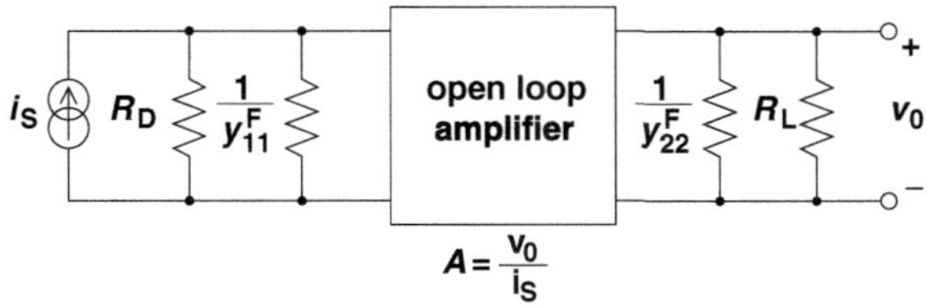

Fig. 6.7. A-circuit for the shunt–shunt feedback amplifier

Then the transresistance of the open-loop amplifier A is calculated from the circuit shown in Fig. 6.7, which includes the loading effects of y_{11}^{F}, y_{22}^{F}, R_D, and R_L.

The gain finally is calculated directly from the A-circuit (Fig. 6.7).

6.1.2 Input and Output Resistance

Input Resistance

The input resistance R_{IN} of the closed loop shunt–shunt feedback amplifier (Fig. 6.2) can be calculated using the two-port description in (6.9) [95]. The input resistance is

$$R_{IN} = \frac{v_1}{i_S}. \tag{6.13}$$

Solution of (6.9) for i_S in terms of v_1 gives

$$i_S = (G_D + y_{11}^{T})v_1 + y_{12}^{F}\frac{-y_{21}^{A}}{(y_{22}^{T} + G_L)}v_1, \tag{6.14}$$

which can be rearranged as

$$R_{IN} = \frac{1}{(G_D + y_{11}^{T})\left[1 + \frac{-y_{21}^{A}}{(G_D + y_{11}^{T})(y_{22}^{T} + G_L)}y_{12}^{F}\right]} = \frac{\left(\frac{1}{G_D + y_{11}^{T}}\right)}{1 + A\beta_F} = \frac{R_{IN}^{A}}{1 + A\beta_F}. \tag{6.15}$$

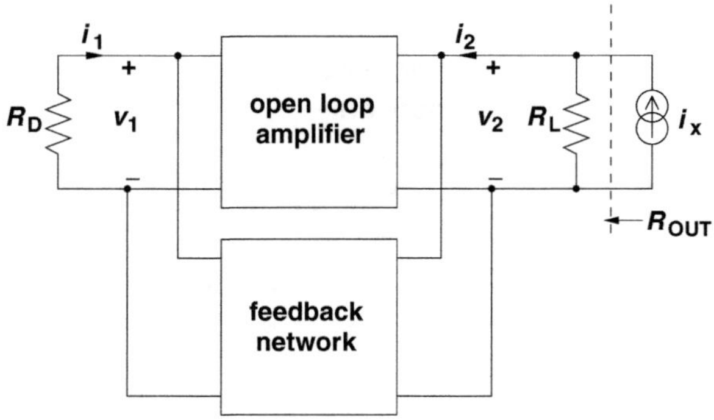

Fig. 6.8. Output resistance of the shunt–shunt feedback amplifier [95]

Shunt feedback reduces the resistance at the port by the factor $(1 + A\beta_F)$. As the loop gain approaches infinity – for an ideal operational amplifier, for example – the input resistance of the closed-loop transresistance amplifier approaches zero. For gigabit optical receivers, however, infinity is not achieved.

Output Resistance

The output resistance of the closed-loop amplifier can be obtained in a similar manner using the circuit shown in Fig. 6.8 [95].

Here we start with (6.7) and apply a test source i_x to the output of the amplifier:

$$\begin{aligned} i_1 &= y_{11}^{\mathrm{T}} v_1 + y_{12}^{\mathrm{F}} v_2, \\ i_2 &= y_{21}^{\mathrm{A}} v_1 + y_{22}^{\mathrm{T}} v_2. \end{aligned} \tag{6.16}$$

The voltage and current at the input port and at the output port are related by

$$\begin{aligned} i_1 &= -G_{\mathrm{D}} v_1, \\ i_2 &= i_{\mathrm{x}} - G_{\mathrm{L}} v_2. \end{aligned} \tag{6.17}$$

Substituting (6.17) into (6.16) results in

$$\begin{aligned} 0 &= (G_{\mathrm{D}} + y_{11}^{\mathrm{T}}) v_1 + y_{12}^{\mathrm{F}} v_x, \\ i_x &= y_{21}^{\mathrm{A}} v_1 + (y_{22}^{\mathrm{T}} + G_{\mathrm{L}}) v_{\mathrm{x}}. \end{aligned} \tag{6.18}$$

When we solve for i_x in terms of v_x, we obtain

$$i_x = y_{21}^{\mathrm{A}} \frac{-y_{12}^{\mathrm{F}}}{(G_{\mathrm{D}} + y_{11}^{\mathrm{T}})} v_x + (y_{22}^{\mathrm{T}} + G_{\mathrm{L}}) v_x. \tag{6.19}$$

Rearranging (6.19) yields the result for the output resistance of the overall amplifier:

$$R_{\mathrm{OUT}} = \frac{v_x}{i_x} = \frac{1}{(y_{22}^{\mathrm{T}} + G_{\mathrm{L}})\left[1 + \frac{-y_{21}^{\mathrm{A}}}{(G_{\mathrm{D}} + y_{11}^{\mathrm{T}})(y_{22}^{\mathrm{T}} + G_{\mathrm{L}})} y_{12}^{\mathrm{F}}\right]}$$

$$= \frac{\left(\frac{1}{y_{22}^{\mathrm{T}} + G_{\mathrm{L}}}\right)}{1 + A\beta_{\mathrm{F}}} = \frac{R_{\mathrm{OUT}}^{\mathrm{A}}}{1 + A\beta_{\mathrm{F}}} . \tag{6.20}$$

The output resistance of the closed-loop amplifier is equal to the output resistance of the A-circuit decreased by the amount of feedback $(1 + A\beta_{\mathrm{F}})$. In the ideal case – approximately achieved by operational amplifiers – the output resistance of the transresistance amplifier approaches zero when the loop gain approaches infinity.

6.2 TIA with Ideal Amplifier

For a TIA, consisting of an ideal amplifier with open-loop gain A_0, a feedback resistor R_{F} and a total input-node capacitance C_{T}, shown in Fig. 6.9, the relationship between the output voltage and the input current is described in (6.25). For the ideal amplifier with constant A_0 over the bandwidth, the feedback capacitance shown in Fig. 6.9 can be neglected, because it is much smaller than C_{T}.

For the circuit in Fig. 6.9 the transimpedance $v_{\mathrm{out}}/i_{\mathrm{in}}$ is derived by (6.21)–(6.24).

The input current is split into a current through the input-node capacitance i_{C} and a current through the feedback resistor i_{R}:

$$i_{\mathrm{in}} = i_{\mathrm{R}} + i_{\mathrm{C}}. \tag{6.21}$$

The input node voltage v_{in} is defined by

$$v_{\mathrm{in}} = \frac{i_{\mathrm{C}}}{j\omega C_{\mathrm{T}}}. \tag{6.22}$$

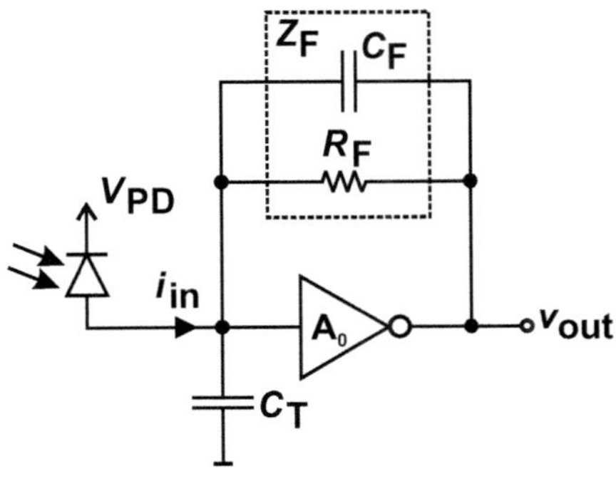

Fig. 6.9. Block diagram of receiver front-end

This input node voltage is amplified to the output node voltage v_{out} by the ideal inverting amplifier with the gain A_0

$$v_{\text{out}} = -v_{\text{in}} A_0 ; \tag{6.23}$$

furthermore it is considered that

$$v_{\text{in}} - v_{\text{out}} = i_{\text{R}} R . \tag{6.24}$$

Inserting (6.23) into (6.24) and calculating i_{C} from (6.22), inserting it into (6.21), i_{R} can be derived in dependence of i_{in} and v_{out}. Substituting i_{R} in (6.24) and deriving $\frac{v_{\text{out}}}{i_{\text{in}}}$ leads to (6.25) [96].

$$\frac{v_{\text{out}}}{i_{\text{in}}} = \frac{-A_0}{A_0+1} \frac{R_{\text{F}}}{1 + \text{j}\omega C_{\text{T}} \frac{R_{\text{F}}}{A_0+1}} . \tag{6.25}$$

The total input node capacitance includes the capacitance of the photodiode C_{pd}, the capacitance of the input transistors C_{in} of the amplifier and parasitic capacitances C_{para}, for example input-pad capacitance, the capacitance of the output pad of the photodiode or of ESD-protection, respectively.

$$C_{\text{T}} = C_{\text{pd}} + C_{\text{para}} + C_{\text{in}} . \tag{6.26}$$

The $-3\,\text{dB}$ bandwidth $f_{-3\,\text{dB}}$ is conditioned by the total input-node capacitance, the open-loop gain A_0 and the feedback resistor R_{F}.

$$f_{-3\text{dB}} = \frac{A_0+1}{2\pi R_{\text{F}} C_{\text{T}}} . \tag{6.27}$$

To increase $f_{-3\,\text{dB}}$ it is possible to enlarge the open-loop gain A_0, or to lower the feedback resistor R_{F} or the input-node capacitance C_{T}, respectively.

6.3 TIA with Frequency-Dependent Open-Loop Gain

A three-stage amplifier is approximately a third-order system

$$A(f) = \frac{A_0}{\left(1 + \text{j}A_0 \frac{f}{f_{\text{g1}}}\right)\left(1 + \text{j}A_0 \frac{f}{f_{\text{g2}}}\right)\left(1 + \text{j}A_0 \frac{f}{f_{\text{g3}}}\right)} \tag{6.28}$$

with three different cut-off frequencies f_{g1}, f_{g2}, and f_{g3}. Equation (6.28) shows the open-loop gain $A(f)$ of a three-stage amplifier. This leads to a relationship between the output voltage and the input current of the transimpedance amplifier as described in (6.29).

$$A_{\text{TIA}}(f) = \frac{v_{\text{out}}(f)}{i_{\text{in}}(f)} = \frac{-A(f)}{A(f)+1} \frac{Z_{\text{F}}}{1 + \text{j}\omega C_{\text{T}} \frac{Z_{\text{F}}}{A(f)+1}} \tag{6.29}$$

where Z_{f} comes from the feedback network consisting of the feedback resistor R_{f} and a feedback capacitance C_{f} (see Fig. 6.9). This time, the feedback

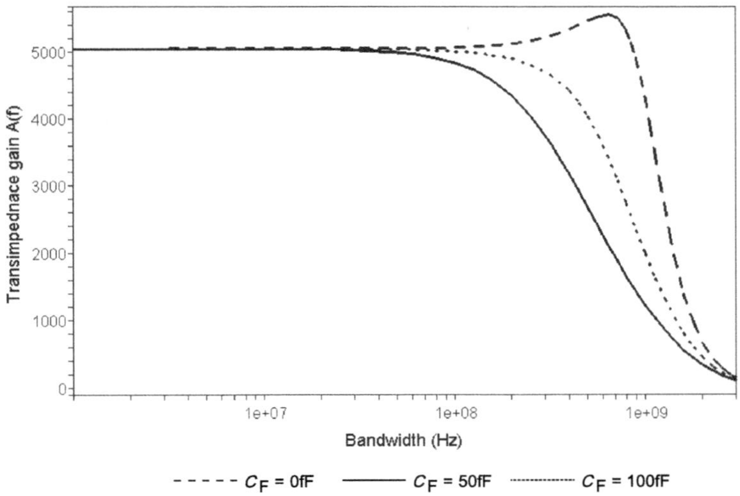

Fig. 6.10. Effective transimpedance of three-stage system with variable feedback capacitance

capacitance cannot be neglected, because it influences the gain peaking for the system at high frequencies. The gain peaking arises from the frequency-dependent open-loop gain of the amplifier.

Inserting (6.28) and $Z_{\mathrm{F}} = \frac{R_{\mathrm{F}}}{1+\mathrm{j}R_{\mathrm{F}}C_{\mathrm{F}}}$ gives the frequency dependence of the transimpedance (see Fig. 6.10). The low-frequency open-loop gain A_0 equals -100, the feedback resistor R_{F} is $5\,\mathrm{k\Omega}$ and the input-node capacitance is assumed with $C_{\mathrm{T}} = 2\,\mathrm{pF}$ and the cut-off frequencies of the amplifier are assumed with $f_{\mathrm{g1}} = 0.5\,\mathrm{GHz}$, $f_{\mathrm{g2}} = 0.9\,\mathrm{GHz}$, and $f_{\mathrm{g3}} = 0.6\,\mathrm{GHz}$. Figure 6.10 depicts the ac response for different feedback capacitances C_{F} of 100 and 50 fF and without C_{F} ($C_{\mathrm{F}} = 0$).

It can be clearly seen that the feedback capacitance C_{F} influences the high-frequency gain of the TIA and therefore can avoid gain peaking for high input-node capacitances.

6.3.1 TIA with Folded-Cascode Amplifier Stage

The folded-cascode TIA circuit is shown in Fig. 6.11. The input transistor M1 has a large width to get high gain and good noise performance; the transistor M2 is the cascode stage, giving low impedance for the drain of M1 and reducing the effective drain-gate capacitance (Miller capacitance of M1). The transistors M3 and M4 are current sources [97, 98]. The output node is of high impedance, therefore a source follower (M5, M6) is used to take care of the load.

The mesh and node equations for the folded-cascode amplifier consisting of the transistors M1 to M4 in Fig. 6.11 are given in (6.31)–(6.39). The small-signal equivalent circuit for the folded-cascode circuit is shown in Fig. 6.12. The small-signal output resistance of each transistor is represented by $r_{\mathrm{o}x}$,

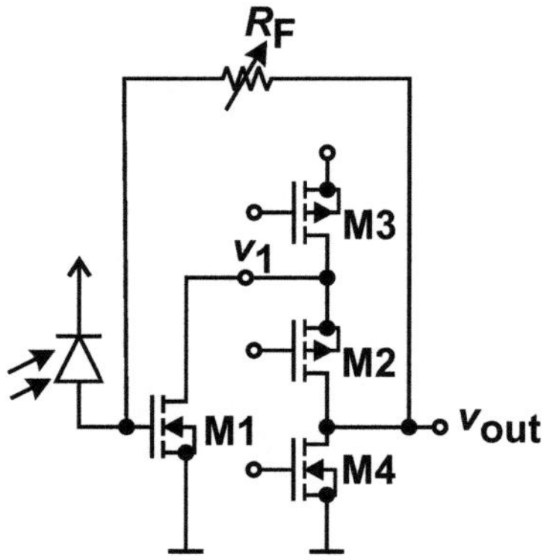

Fig. 6.11. Folded-cascode TIA circuit

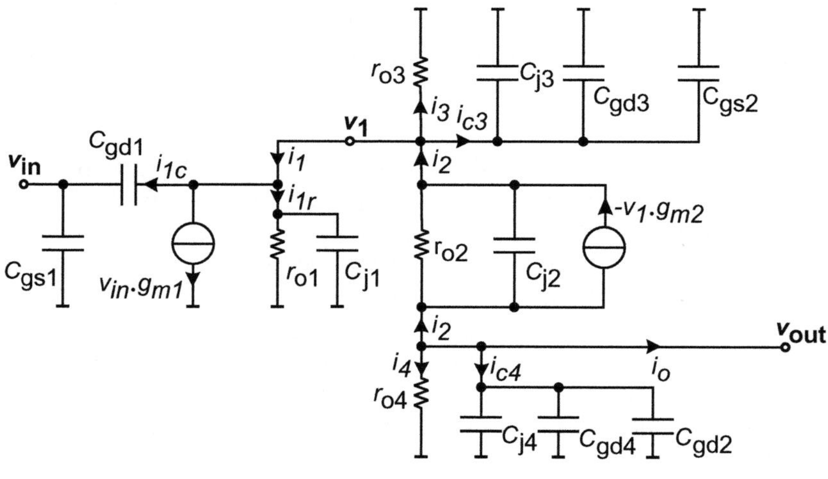

Fig. 6.12. Small-signal equivalent circuit of folded-cascode circuit

gate–source C_{gsx} and gate–drain capacitances C_{gdx} and the drain-well or junction capacitances C_{jx} are included in the small-signal equivalent circuit for each transistor.

$$v_1 - v_{in} = \frac{i_{1c}}{j2\pi f C_{gd1}} \tag{6.30}$$

$$v_1 = r_{o3} i_3 \tag{6.31}$$

where v_1 is the signal voltage at the drain of the M_1. The gate–source capacitances of the transistors C_{gsx} and the drain-gate capacitances C_{gdx} together with the drain-well or junction capacitances C_{jx} are the main parasitic capacitances of the network

$$v_1 = \frac{i_{1r}}{j2\pi f C_{j1} + \frac{1}{r_{o1}}}. \tag{6.32}$$

Transistor M_3 is used as current source and leads with C_{gs2} added to the capacitances of M_3 to

$$v_1 = \frac{i_{3c}}{j2\pi f\left(C_{j3} + C_{gd3} + C_{gs2}\right)}. \tag{6.33}$$

The cascode transistor M_2 gives the following equation for the current through M_2:

$$i_2 = -v_1 g_{m2} + \frac{\dfrac{v_{out} - v_1}{r_{o2}}}{1 + j2\pi f C_{j2} r_{o2}}, \tag{6.34}$$

which leads with (6.36) to the node equation at the drain of M_1

$$i_1 = i_{1c} + i_{1r} + g_{m1} v_{in}, \tag{6.35}$$

$$i_2 = i_1 + i_3 + i_{c3}. \tag{6.36}$$

Equivalently the output current i_o is given by

$$-i_o = i_2 + i_4 + i_{4c}, \tag{6.37}$$

which leads with the current source M_4 to

$$v_{out} = i_4 r_{o4} \tag{6.38}$$

and

$$v_{out} = \frac{i_{4c}}{j2\pi f\left(C_{j4} + C_{gd4} + C_{gd2}\right)}. \tag{6.39}$$

These mesh and node equations enable the calculation of the frequency dependent open-loop gain of the folded-cascode TIA.

The diagram in Fig. 6.13 shows the open-loop gain $A(f) = v_{out}/v_{in}$ of the folded-cascode TIA according to the given (6.30)–(6.39) solved for the open-loop gain of the folded cascode. The curve is derived in Maple with the data summarized in Table 6.1.

All these values will also be used for the following calculations concerning the folded-cascode TIA.

For $f = 0$ the open-loop gain of the folded-cascode TIA is given in (6.40).

$$A_{FC_0} = -\frac{\left(1 + g_{m2}\, r_{o2}\right) r_{o1}\, g_{m1}\, r_{o4}\, r_{o3}}{r_{o1}\, r_{o3}\, g_{m2}\, r_{o2} + r_{o1}\, r_{o3} + r_{o3}\, r_{o4} + r_{o1}\, r_{o4} + r_{o1}\, r_{o2} + r_{o3}\, r_{o2}}. \tag{6.40}$$

This equation results in a gain of the folded-cascode circuit $A_0 = -45.48$ for the values given above.

The gain of the folded-cascode circuit is inserted into (6.25) for the TIA circuit. This leads to the following mesh and node equations (6.41)–(6.44). The input-node capacitance consists mainly of the capacitance of the photodiode C_{pd}, parasitic capacitances C_{para} and the gate capacitance of the input

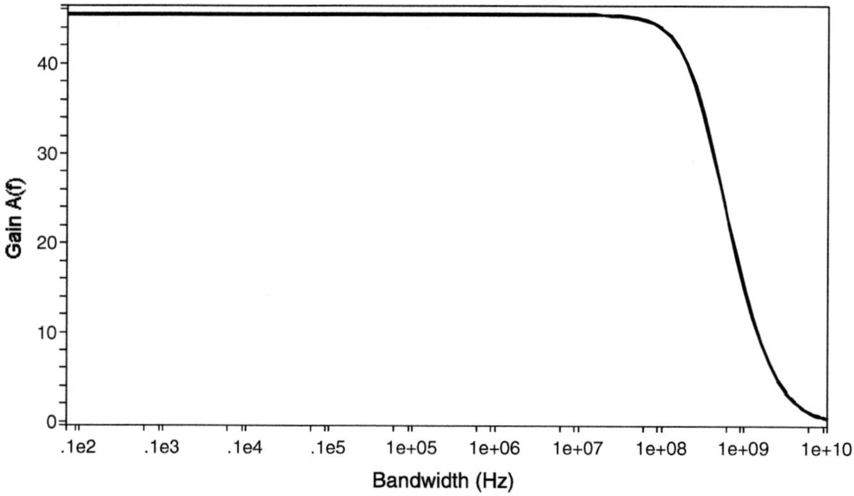

Fig. 6.13. Frequency-dependent gain of the folded-cascode

Table 6.1. Transistor data belonging to Fig. 6.12

transcond. (mS)	output res. (Ω)	junction cap. (fF)	gate–source cap. (fF)	gate–drain cap. (fF)
$g_{m1} = 46.15$	$r_{o1} = 703$	$C_{j1} = 81$	$C_{gs1} = 415.8$	$C_{gd1} = 205$
$g_{m2} = 26.41$	$r_{o2} = 1006$	$C_{j2} = 108$	$C_{gs2} = 554.4$	$C_{gd2} = 274$
	$r_{o3} = 222$	$C_{j3} = 252$	$C_{gs3} = 1290$	$C_{gd3} = 640$
	$r_{o4} = 1520$	$C_{j4} = 7.2$	$C_{gs4} = 36.9$	$C_{gd4} = 18.3$

transistor M_1 (see (6.26)). The parasitic capacitance C_{para} is approximately 100 fF and the photodiode capacitance is about C_{pd} =1.2 pF.

Figure 6.14 shows the small-signal equivalent circuit for the whole folded-cascode TIA. It considers the load of the following stage, represented by the load capacitance C_{load}.

The currents at the input node lead to

$$i_{\mathrm{in}} = i_{\mathrm{c}} + i_{\mathrm{r}} - i_{1\mathrm{c}} \tag{6.41}$$

with the incoming current from the photodiode i_{in}, the current via the gate–drain capacitance of M1 $i_{1\mathrm{c}}$, and the currents through the input node capacitance i_{c} and through the feedback resistance i_{r}.

The output node leads to (6.42) for the load current where i_{o} is the output current of the folded-cascode amplifier

$$i_{\mathrm{load}} = i_{\mathrm{r}} + i_{\mathrm{o}}. \tag{6.42}$$

The "outside" mesh equations for the whole TIA are given by

$$\frac{i_{\mathrm{c}}}{\mathrm{j}2\pi f C_{\mathrm{T}}} = v_{\mathrm{in}} \tag{6.43}$$

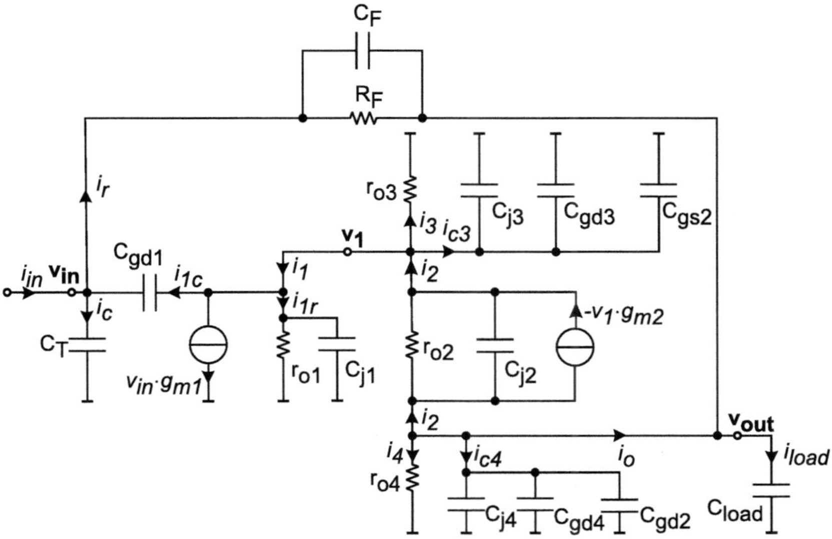

Fig. 6.14. Small-signal equivalent circuit of folded-cascode TIA

with the input node capacitance C_T as given in (6.26), C_F is the feedback capacitance parallel to the feedback resistor R_F,

$$(v_\mathrm{in} - v_\mathrm{out}) = i_\mathrm{r} \frac{R_\mathrm{F}}{1 + \mathrm{j}2\pi f C_\mathrm{F} R_\mathrm{F}}, \tag{6.44}$$

$$v_\mathrm{out} = \frac{i_\mathrm{load}}{\mathrm{j}2\pi f C_\mathrm{load}} \tag{6.45}$$

and the load capacitance $C_\mathrm{load} = 50\,\mathrm{fF}$ as input capacitance of the following circuit.

The "inner" equations of the folded-cascode amplifier itself are given in (6.31)–(6.39).

Calculating the effective transimpedance $A_\mathrm{TIA}(f) = v_\mathrm{out}/i_\mathrm{in}$ and inserting the data of Table 6.1 and $C_\mathrm{F} = 80\,\mathrm{fF}$ leads to the frequency response shown in Fig. 6.15.

This leads to an effective low-frequency transimpedance of $A_\mathrm{TIA,0}$ of $-1761.3\,\Omega$ with a feedback resistor $R_\mathrm{F} = 1.8\,\mathrm{k}\Omega$ and a $-3\,\mathrm{dB}$ cut-off frequency $f_{-3\,\mathrm{dB}} \approx 850\,\mathrm{MHz}$.

6.3.2 TIA with Inverter Amplifier Stages

Figure 6.16 shows the schematic of a three-inverter TIA with the three inverter stages P1/N1, P2/N2, and P3/N3, respectively. The transistors D1 to D3 are so-called diode loads for each inverter stage.

To calculate the transfer function of the TIA, the small-signal equivalent circuit of the three-inverter TIA is on closer examination, see Fig. 6.17.

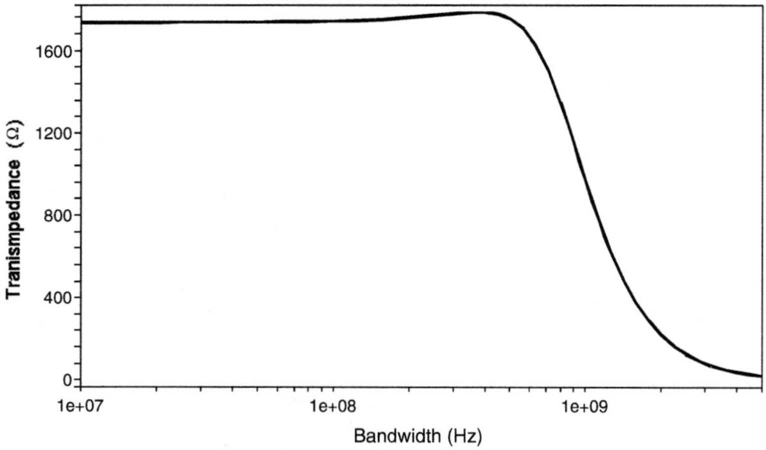

Fig. 6.15. Frequency dependence of the effective transimpedance A_{TIA} of folded-cascode TIA

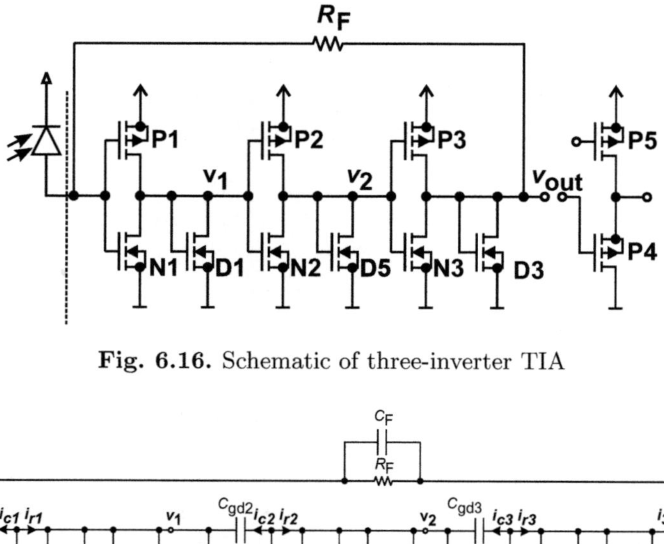

Fig. 6.16. Schematic of three-inverter TIA

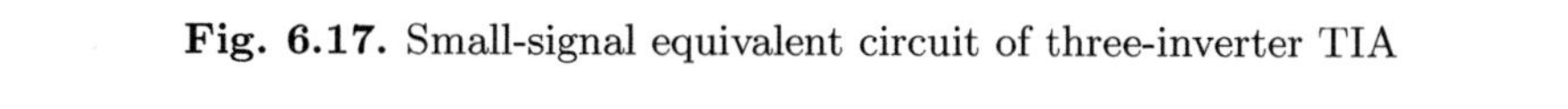

Fig. 6.17. Small-signal equivalent circuit of three-inverter TIA

In this first estimation the transistor model only consists of the transconductance g_{m}, the gate–source capacitance C_{gs} and the drain-well or junction capacitance C_{j}. This very simple model neglects all other capacitances, e.g., the gate–drain capacitance.

The data for the transistors shown in Fig. 6.16 are listed in Table 6.2.

Table 6.2. Transistor data belonging to Fig. 6.16

transcond. (mS)	output admit. (mS)	junction cap. (fF)	gate–source cap. (fF)	gate–drain cap. (fF)
$g_{\mathrm{p1}} = 61.8$	$g_{\mathrm{dsp1}} = 5.5$	$C_{\mathrm{jp1}} = 31.2$	$C_{\mathrm{gsp1}} = 158.9$	$C_{\mathrm{gdp1}} = 158.9$
$g_{\mathrm{n1}} = 122$	$g_{\mathrm{dsn1}} = 8.7$	$C_{\mathrm{jn1}} = 42.2$	$C_{\mathrm{gsn1}} = 216.8$	$C_{\mathrm{gdn1}} = 216.8$
$g_{\mathrm{p2}} = 32.9$	$g_{\mathrm{dsp2}} = 2.85$	$C_{\mathrm{jp2}} = 15.6$	$C_{\mathrm{gsp2}} = 76.4$	$C_{\mathrm{gdp2}} = 76.4$
$g_{\mathrm{n2}} = 58.7$	$g_{\mathrm{dsn2}} = 2.32$	$C_{\mathrm{jn2}} = 31.7$	$C_{\mathrm{gsn2}} = 162.6$	$C_{\mathrm{gdn2}} = 76.4$
$g_{\mathrm{p3}} = 32.8$	$g_{\mathrm{dsp3}} = 2.9$	$C_{\mathrm{jp3}} = 15.6$	$C_{\mathrm{gsp3}} = 76.4$	$C_{\mathrm{gdp2}} = 76.4$
$g_{\mathrm{n3}} = 54.1$	$g_{\mathrm{dsn3}} = 2.0$	$C_{\mathrm{jn3}} = 31.7$	$C_{\mathrm{gsn3}} = 162.6$	$C_{\mathrm{gdn2}} = 76.4$
$g_{\mathrm{d1}} = 8.9$	$g_{\mathrm{dsd1}} = 0.03$	$C_{\mathrm{jd1}} = 2.4$	$C_{\mathrm{gsd1}} = 12.3$	$C_{\mathrm{gdd1}} = 12.3$
$g_{\mathrm{d2}} = 6.0$	$g_{\mathrm{dsd2}} = 0.22$	$C_{\mathrm{jd2}} = 3.6$	$C_{\mathrm{gsd2}} = 18.5$	$C_{\mathrm{gdd2}} = 18.5$
$g_{\mathrm{d3}} = 19.9$	$g_{\mathrm{dsd3}} = 0.73$	$C_{\mathrm{jd3}} = 11.7$	$C_{\mathrm{gsd3}} = 60.1$	$C_{\mathrm{gdd2}} = 60.1$

Table 6.3. Transistor data belonging to Fig. 6.17

transcond. (mS)	output admit. (mS)	junction cap. (fF)	gate–source cap. (fF)	gate–drain cap. (fF)
$g_{\mathrm{m1}}^{\mathrm{I}} = 183.8$	$g_{\mathrm{ds1}} = 5.5$	$C_{\mathrm{j1}} = 73.4$	$C_{\mathrm{gs1}} = 375.7$	$C_{\mathrm{gdp1}} = 357.7$
$g_{\mathrm{m2}}^{\mathrm{I}} = 91.6$	$g_{\mathrm{ds2}} = 5.17$	$C_{\mathrm{j2}} = 47.3$	$C_{\mathrm{gs2}} = 239.0$	$C_{\mathrm{gd2}} = 239.0$
$g_{\mathrm{m3}}^{\mathrm{I}} = 86.9$	$g_{\mathrm{ds3}} = 4.9$	$C_{\mathrm{j3}} = 47.3$	$C_{\mathrm{gs3}} = 239.0$	$C_{\mathrm{gdp2}} = 239.0$
$g_{\mathrm{d1}} = 8.9$	$g_{\mathrm{dsd1}} = 0.03$	$C_{\mathrm{jd1}} = 2.4$	$C_{\mathrm{gsd1}} = 12.3$	$C_{\mathrm{gdd1}} = 12.3$
$g_{\mathrm{d2}} = 6.0$	$g_{\mathrm{dsd2}} = 0.22$	$C_{\mathrm{jd2}} = 3.6$	$C_{\mathrm{gsd2}} = 18.5$	$C_{\mathrm{gdd2}} = 18.5$
$g_{\mathrm{d3}} = 19.9$	$g_{\mathrm{dsd3}} = 0.73$	$C_{\mathrm{jd3}} = 11.7$	$C_{\mathrm{gsd3}} = 60.1$	$C_{\mathrm{gdd2}} = 60.1$

The small-signal equivalent circuit is already simplified. The drawn capacitances C_{gs1}, C_{gs2}, C_{gs3} represent the gate–source capacitances of both of the inverter transistors. For example, the capacitance $C_{\mathrm{gs1}} = C_{\mathrm{gsn1}} + C_{\mathrm{gsp1}}$ represents the sum of the gate–source capacitances of P1 C_{gsp1} and N1 C_{gsn1}. The junction capacitances C_{j1}, C_{j2}, and C_{j3} also represent the sum of PMOS and NMOS of each inverter. Also the transconductances $g_{\mathrm{x}}^{\mathrm{I}}$ and the output resistance of the inverter stages r_{ox} summarize both transistors of the inverter, where the x stands for the number of the stage.

For the diode loads the conductance $g_{\mathrm{d}x}$ summarizes the transconductance $g_{\mathrm{dn}x}$ and the output resistance $r_{\mathrm{od}x}$ of the diode transistor Dx. The capacitances $C_{\mathrm{d}x}$ summarize the junction capacitance and the gate–source capacitance of the load transistors. Again the x stands for the number of the stage.

The small-signal equivalent circuit shown in Fig. 6.17 contains the combined values for the PMOS and NMOS transistors of the inverters. These values according to Fig. 6.17 are summarized in Table 6.3. This table also shows values for the gate–drain capacitances, which will be used later.

The transfer function of each inverter stage with diode load is calculated as follows:

$$A_x = -\frac{v_{x+1}}{v_\mathrm{x}} = -\frac{g_x^I}{\frac{1}{r_{0x}} + \mathrm{j}2\pi f\left(C_{\mathrm{j}x} + C_{\mathrm{d}x} + C_{\mathrm{gs}(x+1)}\right) + g_{\mathrm{d}x}}. \tag{6.46}$$

This means for the three inverter stages of Fig. 6.17:

$$A_1 = -\frac{v_1}{v_\mathrm{in}} = -\frac{g_\mathrm{m1}^I}{\frac{1}{r_{01}} + \mathrm{j}2\pi f\left(C_\mathrm{j1} + C_\mathrm{d1} + C_\mathrm{gs2}\right) + g_\mathrm{d1}}. \tag{6.47}$$

Calculated with the values of Table 6.3 this leads to a low-frequency gain of the first stage $A_{10} = -7.93$ and a $-3\,\mathrm{dB}$ cut-off frequency of $f_{-3\mathrm{dB}} \approx 10\,\mathrm{GHz}$.

The low-frequency gain of the second stage

$$A_2 = -\frac{v_2}{v_1} = -\frac{g_\mathrm{m2}^I}{\frac{1}{r_{02}} + \mathrm{j}2\pi f\left(C_\mathrm{j2} + C_\mathrm{d2} + C_\mathrm{gs3}\right) + g_\mathrm{d2}} \tag{6.48}$$

leads to $A_{20} = -8.034$ with a $-3\,\mathrm{dB}$ cut-off frequency of $f_{-3\mathrm{dB}} \approx 5.9\,\mathrm{GHz}$.

The calculations for the third inverter lead to

$$A_3 = -\frac{v_\mathrm{out}}{v_2} = -\frac{g_\mathrm{m3}^\mathrm{I}}{\frac{1}{r_{03}} + \mathrm{j}\,2\pi f\left(C_\mathrm{j3} + C_\mathrm{d3}\right) + g_\mathrm{d3}} \tag{6.49}$$

with a low-frequency gain of $A_{30} = -3.41$ and a $-3\,\mathrm{dB}$ cut-off frequency of $f_{-3\,\mathrm{dB}} \approx 18.7\,\mathrm{GHz}$. The curves for the gain of the single stages are shown in Fig. 6.18.

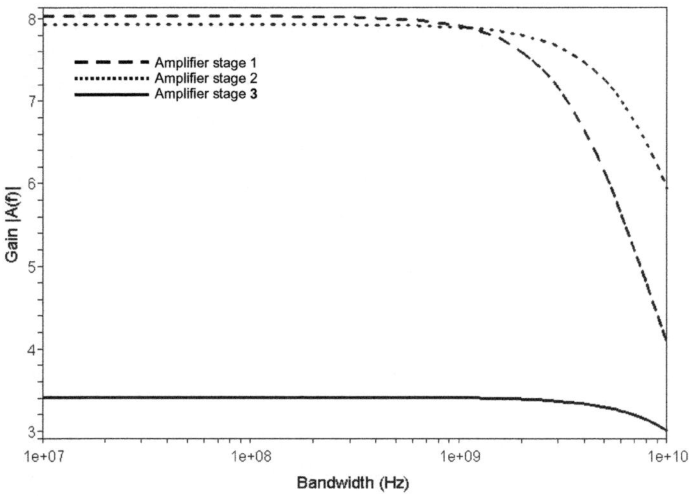

Fig. 6.18. Frequency dependence of the absolute value of the single-stage gains, for stage 1 to stage 3

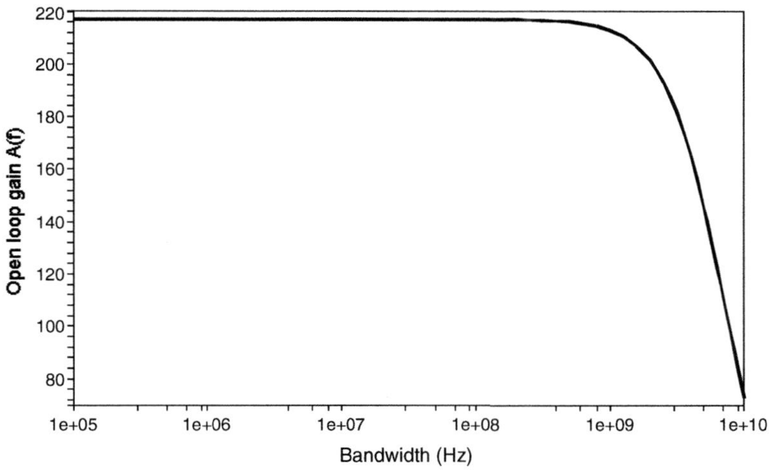

Fig. 6.19. Open-loop gain of three-inverter TIA

The open-loop gain of the three-inverter TIA $A_{3\text{I}}$ is calculated by multiplying the single stage gains

$$A_{3\text{I}} = A_1 A_2 A_3. \tag{6.50}$$

Inserting the values of Table 6.3 into (6.50) gives

$$A_{3\text{I}}(f) = \frac{v_{\text{out}}}{v_{\text{in}}} = -\frac{g_{\text{m}1}^{I} g_{\text{m}2}^{I} g_{\text{m}3}^{I}}{\left(\frac{1}{r_{01}} + \text{j}2\pi f\left(C_{\text{j}1} + C_{\text{d}1} + C_{\text{gs}2}\right) + g_{\text{d}1}\right)} \cdot \frac{1}{\left(\frac{1}{r_{02}} + \text{j}2\pi f\left(C_{\text{j}2} + C_{\text{d}2} + C_{\text{gs}3}\right) + g_{\text{d}2}\right)\left(\frac{1}{r_{03}} + \text{j}2\pi f\left(C_{\text{j}3} + C_{\text{d}3}\right) + g_{\text{d}3}\right)}. \tag{6.51}$$

Inserting the values of Table 6.3 into (6.51) results in the frequency-dependent characteristic of the absolute value of the open-loop gain of the TIA shown in Fig. 6.19, with a low-frequency open-loop gain $A_{3\text{I}0} = -217.24$ and a $-3\,\text{dB}$ cut-off frequency of $f_{-3\text{dB}} \approx 4.9\,\text{GHz}$.

Inserting the gain of the three-inverter stages as open-loop gain of the TIA into (6.25) referring to the simple TIA model of Fig. 6.9 leads to an effective transimpedance of the TIA like that shown in Fig. 6.20.

For a feedback resistor $R_{\text{F}} = 5.4\,\text{k}\Omega$ the effective low-frequency transimpedance is $5.375\,\text{k}\Omega$ with a $-3\,\text{dB}$ cut-off frequency of $f_{-3\text{dB}} \approx 5.9\,\text{GHz}$ due to the raising of the transimpedance at about 1 GHz.

It can be clearly seen that this simple model does not meet reality at high frequencies. The low-frequency behavior is quite exact, but for high frequencies it is necessary to use a more complicated model.

Therefore the transistor model is adjusted by the gate–drain capacitances C_{gd}. The next step is to create a small-signal equivalent circuit of the whole

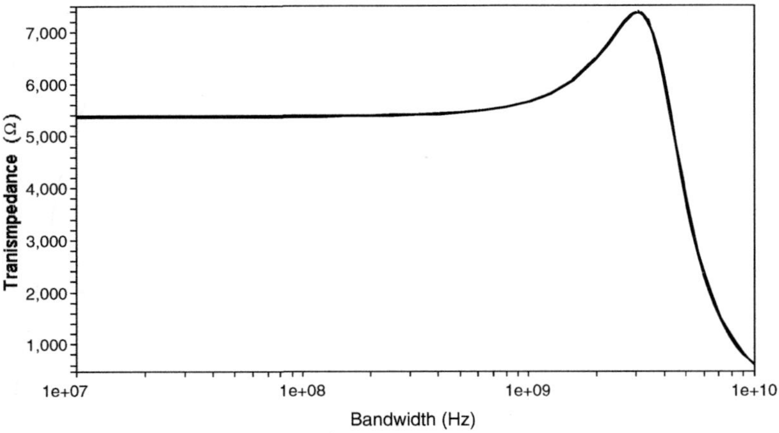

Fig. 6.20. Effective transimpedance of three-inverter TIA (simple model)

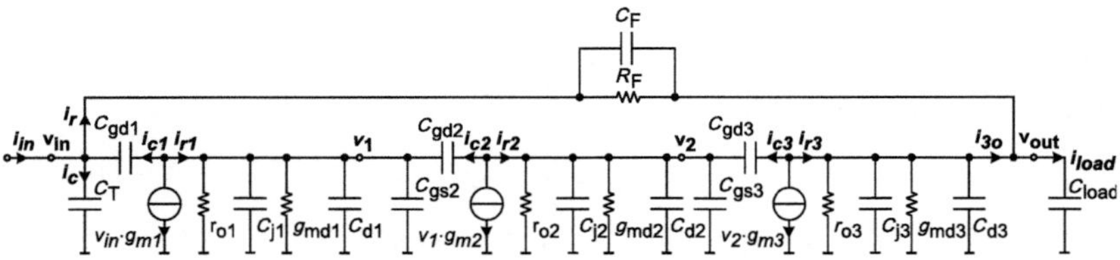

Fig. 6.21. Small-signal equivalent circuit of three-inverter TIA

TIA, not only of the amplifier stages, but to include the load of the following stage correctly. Figure 6.21 shows the equivalent circuit. The model is also enlarged by a feedback capacitance C_F which is about 10 fF. Some of the designs presented later use a PMOS for the feedback resistor and therefore take the parasitic capacitance of the transistor as C_F.

The node equations of the circuit in Fig. 6.21 are shown in (6.52)–(6.56).

The input node currents result in

$$i_{in} + i_{c1} = i_r + i_c \tag{6.52}$$

with the current i_{c1} through the gate–drain capacitance of the input inverter (PMOS and NMOS). The current i_c is the one through the input-node capacitance C_T and i_r flows through the feedback resistance and feedback capacitance.

The following three equations define the currents of the nodes of each inverter stage, where i_{cx} is again the current through C_{gdx}.

$$-v_{in} g_{m1}^{I} = i_{r1} + i_{c1}, \tag{6.53}$$

$$-v_1 g_{m2}^{I} = i_{r2} + i_{c2}, \tag{6.54}$$

$$-v_2 g_{\mathrm{m3}}^{\mathrm{I}} = i_{\mathrm{r3}} + i_{\mathrm{c3}}. \tag{6.55}$$

The output node currents lead to

$$i_{3\mathrm{o}} + i_{\mathrm{r}} = i_{\mathrm{load}} \tag{6.56}$$

where $i_{3\mathrm{o}}$ is the current which loads the third inverter stage.

Inserting the gate–drain capacitances leads to the following equations:

$$(v_1 - v_{\mathrm{in}})\mathrm{j}2\pi f C_{\mathrm{gd1}} = i_{\mathrm{c1}}, \tag{6.57}$$

$$(v_2 - v_1)\mathrm{j}2\pi f C_{\mathrm{gd2}} = i_{\mathrm{c2}}, \tag{6.58}$$

$$(v_{\mathrm{out}} - v_2)\mathrm{j}2\pi f C_{\mathrm{gd3}} = i_{\mathrm{c3}}. \tag{6.59}$$

The "inner" mesh equations of the amplifier are given in (6.60)–(6.62).

$$v_1 = \frac{i_{\mathrm{r1}} + i_{\mathrm{c2}}}{\frac{1}{r_{01}} + \mathrm{j}2\pi f\left(C_{\mathrm{j1}} + C_{\mathrm{d1}} + C_{\mathrm{gs2}}\right) + g_{\mathrm{md1}}} \tag{6.60}$$

$$v_2 = \frac{i_{\mathrm{r2}} + i_{\mathrm{c3}}}{\frac{1}{r_{02}} + \mathrm{j}2\pi f\left(C_{\mathrm{j2}} + C_{\mathrm{d2}} + C_{\mathrm{gs3}}\right) + g_{\mathrm{md2}}} \tag{6.61}$$

$$v_{\mathrm{out}} = \frac{i_{\mathrm{r3}} - i_{3\mathrm{o}}}{\frac{1}{r_{03}} + \mathrm{j}2\pi f\left(C_{\mathrm{j3}} + C_{\mathrm{d3}}\right) + g_{\mathrm{md3}}} \tag{6.62}$$

The load of the TIA stage is considered by (6.63)

$$v_{\mathrm{out}} = \frac{i_{\mathrm{load}}}{\mathrm{j}2\pi f C_{\mathrm{load}}}. \tag{6.63}$$

For the TIA itself the mesh equation over the feedback resistor must be given,

$$v_{\mathrm{in}} = i_{\mathrm{r}} \frac{R_{\mathrm{F}}}{1 + \mathrm{j}2\pi f C_{\mathrm{F}} R_{\mathrm{F}}} \tag{6.64}$$

as well as the connection between input voltage v_{in} and the input current i_{in} given by (6.52) and (6.65)

$$v_{\mathrm{in}} = \frac{i_{\mathrm{c}}}{\mathrm{j}2\pi f C_{\mathrm{T}}} \tag{6.65}$$

Solving (6.52)–(6.65) to get the effective transimpedance $A_{3\mathrm{I}} = v_{\mathrm{out}}/i_{\mathrm{in}}$ leads to a complex formula, but inserting the values of Table 6.3 leads to the frequency-dependent effective transimpedance shown in Fig. 6.22.

This more complicated model leads again to an effective low-frequency transimpedance of $A_{3\mathrm{I}0} = -5375\,\Omega$, but to a $-3\,\mathrm{dB}$ cut-off frequency $f_{-3\,\mathrm{dB}} = 1.35\,\mathrm{GHz}$ which meets reality quite well.

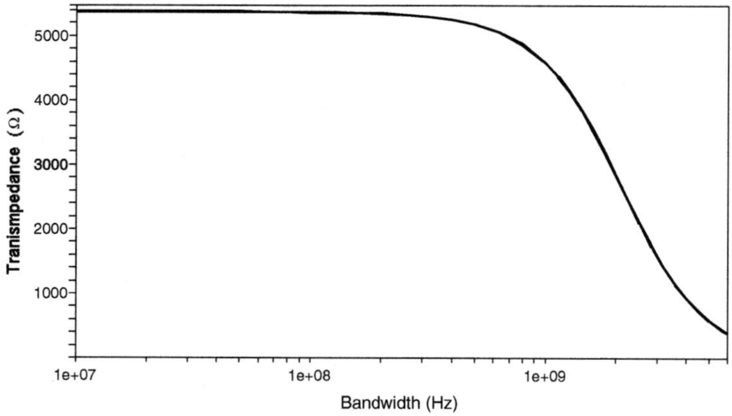

Fig. 6.22. Effective transimpedance of three-inverter TIA

6.3.3 Transimpedance-Gain Switching and Stability of TIA with Inverter Amplifier Stages

The small-signal analysis described before is for the case of minimum optical input power and therefore maximum feedback resistance R_F. Burst-mode receivers have to offer a high dynamic range of the input optical power and therefore the feedback resistance has to be lowered to avoid overdrive and pulse width distortion.

To keep the complete TIA stable normally the feedback capacitance C_F is enlarged to keep $R_F C_F$ constant. The problem of switching this capacitance is that in the used deep-sub-micron CMOS technology this switching leads to a deficit in the bandwidth of the TIA, due to the fact that the parasitic capacitances of the switching transistor limit the switchable C_F. To achieve a high dynamic range, R_F is lowered by a factor of, for example 100; this would necessitate switching a hundred times larger C_F than the fixed wired one. In the designs C_F is in the magnitude of around 10 fF what means that capacitances of about 1 pF have to be switched. Therefore a large switching transistor is necessary, which reduces the bandwidth of the complete system.

This is the reason why the three-stage designs presented in Chap. 9 work with a different approach [99]. Even for low optical input power C_F is kept constant and the open-loop gain is reduced instead. We want to emphasize again that not every design in Chap. 9 has a device C_F, because the small values of C_F are covered by the parasitic capacitances of the transistors in the feedback path which physically build the feedback resistance.

Depicting the loop gain A_{loop} in a Bode diagram and observing the phase margin is the first step of the stability analysis.

Therefore the schematic is abstracted as shown in Fig. 6.23. Each inverter stage, consisting of NMOS, PMOS and diode load is carried out with fixed diode load and an additional variable diode load in parallel. This offers the possibility to decrease the gain of the inverter stage by increasing the load at

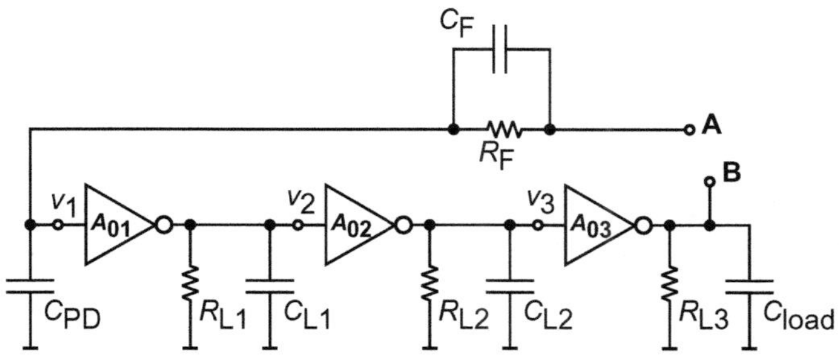

Fig. 6.23. Simplified schematic of the three-inverter TIA and ripping up of the loop

Table 6.4. Device data belonging to Fig. 6.23

transcond. (mS)	load resistance (Ω)	load capacitance (fF)
$g_{\mathrm{m1}} = 183.8$	$R_{\mathrm{L1}} = 69.5$	$C_{\mathrm{L1}} = 315$
$g_{\mathrm{m2}} = 90$	$R_{\mathrm{L2}} = 89.5$	$C_{\mathrm{L2}} = 290$
$g_{\mathrm{m3}} = 86$	$R_{\mathrm{L3}} = 40.3$	$C_{\mathrm{load}} = 30$

the output. Therefore in the following the stages are depicted as an ideal amplifier with the gain $A_{0\mathrm{x}}$ followed by a low-pass filter R_{Lx} and C_{Lx}, where "x" stands for the number of the stage. The gain of the ideal amplifier is given by:

$$A_{0\mathrm{x}} = -g_{\mathrm{mx}} R_{\mathrm{Lx}}, \tag{6.66}$$

which is compatible to the three-stage TIA discussed above. Again g_{mx} represents the sum of the transconductances of NMOS and PMOS of each stage (see Table 6.4). The load resistance R_{Lx} represents the diode load on the one hand and the drain source resistance of the transistors of the inverter in parallel on the other hand. The load capacitances of each stage consist of the gate–source capacitance of the following stage and the junction capacitances of the determined stage. The load capacitance of the whole TIA C_{load} is also the load of the third stage. The feedback resistance R_{F} is 5.4 kΩ like in the example discussed above. The feedback capacitance is assumed with 5 fF. Again the photodiode capacitance is given with 1.2 pF.

For the stability analysis in the Bode diagram the loop is ripped up as shown in Fig. 6.23.

The loop gain is the relation between point A and point B in Fig. 6.23. Figure 6.24 shows the schematic for the calculation of the loop gain A_{loop} of the TIA. To keep the load of the system correct the feedback path is additionally connected to point B in Fig. 6.24.

The cut-off frequency of the first low-pass filter is calculated by

$$\omega_{\mathrm{g1}} = \frac{1}{R_{\mathrm{L1}} C_{\mathrm{L1}}}, \tag{6.67}$$

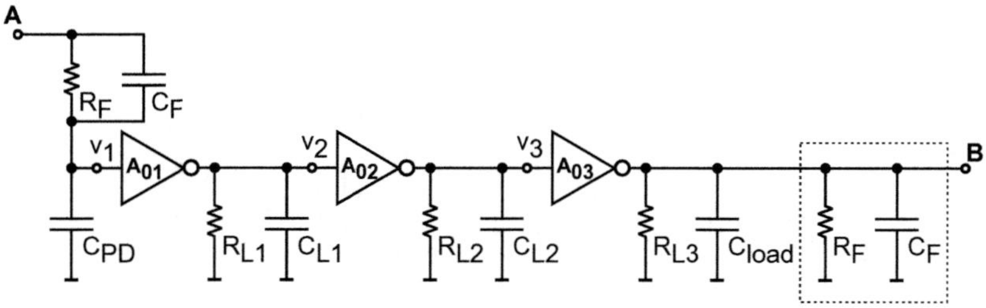

Fig. 6.24. Schematic of ripped up loop

and the cut-off frequency of the second stage is calculated the same way

$$\omega_{\mathrm{g2}} = \frac{1}{R_{\mathrm{L2}} C_{\mathrm{L2}}}. \tag{6.68}$$

The load of the third stage is the sum of the load capacitance and the feedback capacitance. The load resistance is the parallel resistors R_{L3} and R_{f}

$$\omega_{\mathrm{g3}} = \frac{1}{\frac{R_{\mathrm{L3}} R_{\mathrm{f}}}{R_{\mathrm{L3}}+R_{\mathrm{f}}} \left(C_{\mathrm{L3}} + C_{\mathrm{f}}\right)}. \tag{6.69}$$

Together with the $-3\,\mathrm{dB}$ cut-off frequencies and the impedances for the feedback path Z_{F}

$$Z_{\mathrm{F}} = \frac{R_{\mathrm{F}}}{1 + \mathrm{j}\omega C_{\mathrm{F}} R_{\mathrm{F}}} \tag{6.70}$$

and for the photodiode Z_{pd}

$$Z_{\mathrm{pd}} = \frac{1}{\mathrm{j}\omega C_{\mathrm{pd}}} \tag{6.71}$$

the loop gain A_{loop} is calculated as

$$A_{\mathrm{loop}} = \frac{v_{\mathrm{B}}}{v_{\mathrm{A}}} = \frac{Z_{\mathrm{pd}}}{(Z_{\mathrm{F}} + Z_{\mathrm{pd}})} \frac{A_{01}}{\left(1 + \mathrm{j}\frac{\omega}{\omega_{\mathrm{g1}}}\right)} \frac{A_{02}}{\left(1 + \mathrm{j}\frac{\omega}{\omega_{\mathrm{g2}}}\right)} \frac{A_{03}}{\left(1 + \mathrm{j}\frac{\omega}{\omega_{\mathrm{g3}}}\right)} \tag{6.72}$$

For the calculation of the loop gain with the possibility to reduce the feedback resistance R_{F} by a factor F, R_{F}/F is inserted instead of R_{F} in (6.70). The gain of the amplifier will not be reduced in this first estimation to display the requirement of the reduction of the open-loop gain. Inserting (6.66)–(6.71) into (6.72) leads to:

$$A_{\mathrm{loop}} = -\frac{g_{\mathrm{m1}} R_{\mathrm{L1}}\, g_{\mathrm{m2}} R_{\mathrm{L2}}\, g_{\mathrm{m3}} R_{\mathrm{L3}}}{\left(1 + \frac{s R_{\mathrm{L3}} R_{\mathrm{F}} (C_{\mathrm{F}} + C_{\mathrm{load}})}{F\left(R_{\mathrm{L3}} + \frac{R_{\mathrm{f}}}{F}\right)}\right) (1 + s R_{\mathrm{L1}} C_{\mathrm{L1}}) (1 + s R_{\mathrm{L2}} C_{\mathrm{L2}}) \left(s\, C_{\mathrm{PD}} \left(\frac{R_{\mathrm{F}}}{F\left(1 + \frac{C_{\mathrm{F}} R_{\mathrm{F}}}{F}\right)} + \frac{1}{s\, C_{\mathrm{PD}}}\right)\right)}. \tag{6.73}$$

Figure 6.25 depicts the Bode plot of the system with four different R_F values and constant open-loop gain. It can be seen clearly that for small values of R_F the phase is zero while the loop gain is still obviously larger than 0 dB.

Only for the maximum R_F the phase margin is larger than 45° and therefore the phase margin of the system is only large enough for values near the maximum R_F.

To determine the values of the feedback resistance, where the TIA is stable, the root locus is plotted for the closed loop [100]. The loop gain can be used to calculate it easily as shown in Fig. 6.26.

Calculating from input (in) to output (out) leads to

$$\frac{\text{out}}{\text{in}} = \frac{(A_{\text{loop}})}{1 + A_{\text{loop}}} \tag{6.74}$$

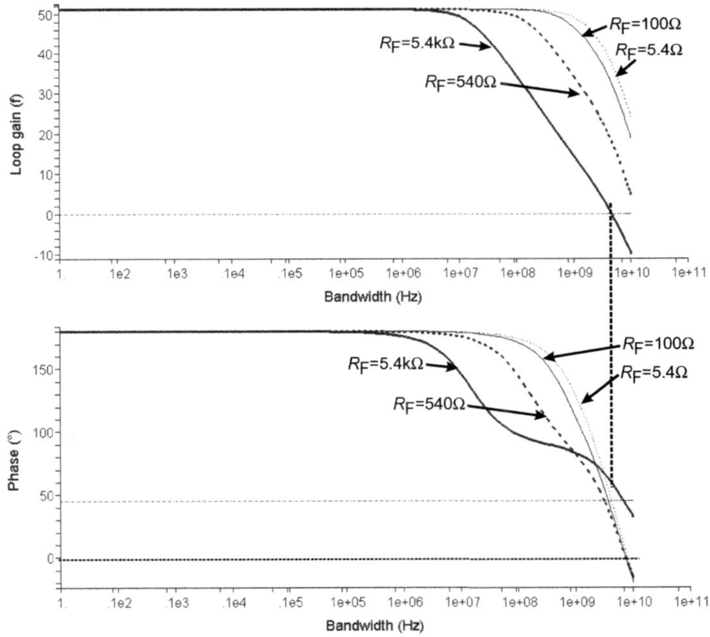

Fig. 6.25. Bode diagram of the loop gain with decreasing feedback resistance

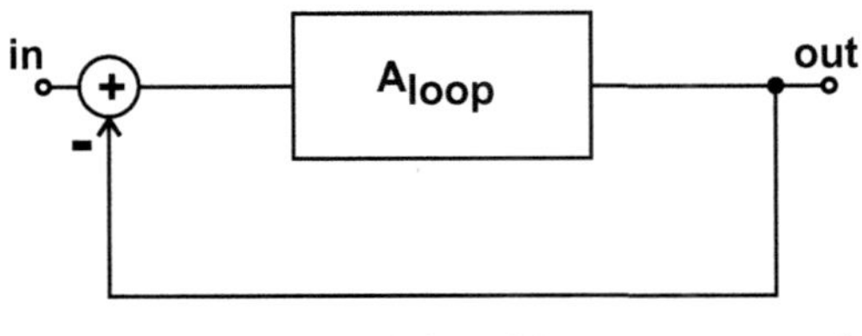

Fig. 6.26. Block diagram of closed loop concerning loop gain

which leads to a characteristic polynomial of

$$P_{\mathrm{char}} = 1 + A_{\mathrm{loop}}. \tag{6.75}$$

The root locus shows that systems with feedback resistors equal or smaller than 1,261Ω have poles in the right-half plane and are therefore unstable (see Fig. 6.27).

Lowering the feedback resistor by a factor requires the lowering of the open-loop gain to avoid this stability problem. Therefore also the load resistances are lowered. This leads to a lowering of the gain, and the cut-off frequency is moved toward higher frequencies. Decreasing R_{F} by a factor of F means also decreasing the load resistances of each stage by $\sqrt[3]{F}$. The loop gain A_{loop} is displayed in (6.76). Compared to (6.73) the load resistance for each stage $R_{\mathrm{L}x}$ is replaced with $\dfrac{R_{\mathrm{L}x}}{\sqrt[3]{F}}$.

Figure 6.28 depicts the Bode plot for four different feedback resistor values.

$$A_{\mathrm{loop}} = - \frac{g_{\mathrm{m1}} R_{\mathrm{L1}}\, g_{\mathrm{m2}} R_{\mathrm{L2}}\, g_{\mathrm{m3}} R_{\mathrm{L3}}}{F \left(1 + \frac{s R_{\mathrm{L3}} R_{\mathrm{F}} (C_{\mathrm{F}} + C_{\mathrm{load}})}{\sqrt[3]{F} F \left(\frac{R_{\mathrm{L3}}}{\sqrt[3]{F}} + \frac{R_{\mathrm{f}}}{F} \right)} \right) \left(1 + \frac{s R_{\mathrm{L1}} C_{\mathrm{L1}}}{\sqrt[3]{F}} \right) \left(1 + \frac{s R_{\mathrm{L2}} C_{\mathrm{L2}}}{\sqrt[3]{F}} \right) \left(s\, C_{\mathrm{PD}} \left(\frac{R_{\mathrm{F}}}{F \left(1 + \frac{C_{\mathrm{F}} R_{\mathrm{F}}}{F} \right)} + \frac{1}{s\, C_{\mathrm{PD}}} \right) \right)}. \tag{6.76}$$

The phase margin for these feedback resistor values is summarized in Table 6.5. With a minimum phase margin of 57° for the largest feedback value

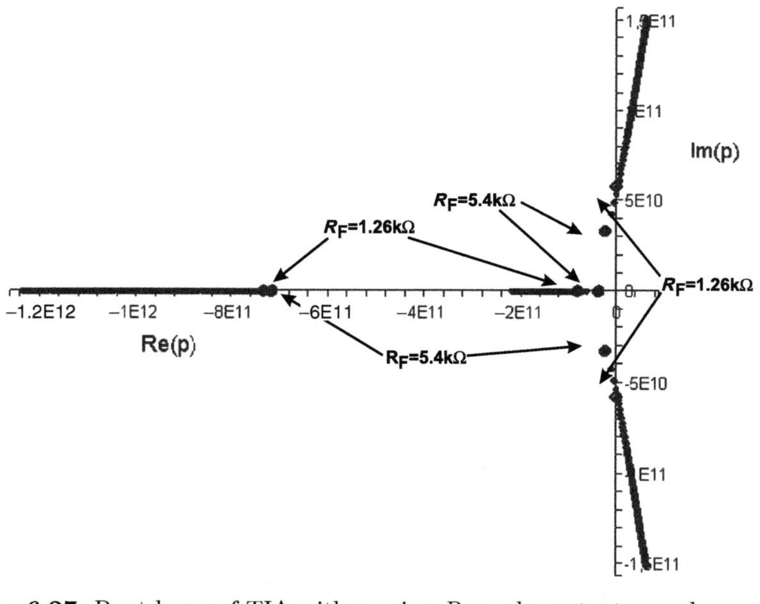

Fig. 6.27. Root locus of TIA with varying R_{F} and constant open-loop gain

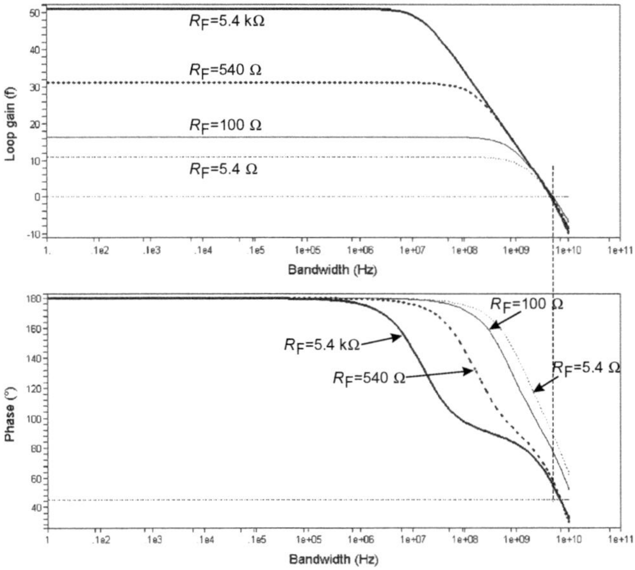

Fig. 6.28. Bode diagram of the loop gain

Table 6.5. Transit frequency and phase margin for different feedback resistances

feedback resistor (Ω)	transit frequency (GHz)	phase margin (°)
5,400	4.5	57.0
540	4.5	58.3
100	5.0	76.7
5.4	5.9	87.9

the circuit is stable. For lower values of the feedback resistor the phase margin is increasing and therefore the stability problem is avoided.

Calculating the poles of the system leads to the root locus dependent on the decreasing factor F. It can be clearly seen in Fig. 6.29 that all poles for all factors F are in the left-half plane and therefore the system is stable for each F.

In the real design the limits of decreasing the feedback resistance and the gain are given by the used devices. Both feedback resistor and diode load are physically formed by transistors. The minimum values are therefore given by the minimum output resistance of the transistors. The diode loads are wider transistors than absolutely necessary for generating the diode load, due to the fact that the current density would be too high for small transistors with a gate voltage of VDD. Therefore the gain is variable in a wider range than the feedback resistance in the most designs described in Chap. 9.

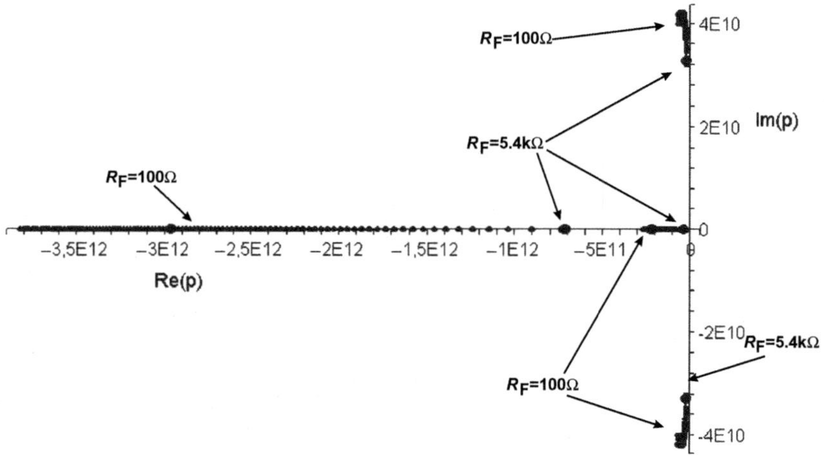

Fig. 6.29. Root locus of TIA with varying R_{F} and varying open-loop gain

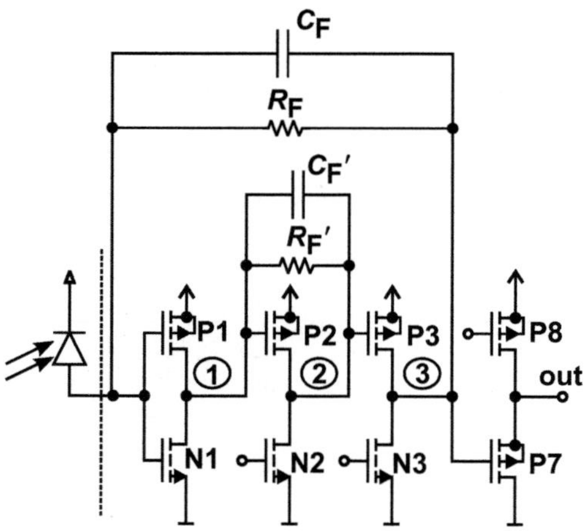

Fig. 6.30. Schematic of three-stage BM TIA with internal feedback

6.3.4 Three-Stage Burst-Mode TIA with Internal Feedback

In Fig. 6.30 another three-stage TIA, this time with internal feedback over the second stage, is depicted. The feedback network is simplified to a single resistor R_{F} and a capacitor C_{F} for a first estimation of the frequency response. C_{F} considers parasitics of the realized feedback network described later.

Table 6.6 summarizes the parameters of the circuit shown in Fig 6.30. The junction capacitances include also estimations of parasitic capacitances due to the layout. The parameters are called like the transistors for example, g_{mn1} for the transconductance of transistor N1.

For the ac analysis of the circuit shown in Fig. 6.30 the small signal equivalent circuit shown in Fig. 6.31 is contemplated.

Table 6.6. Transistor data belonging to Fig. 6.30

transcond. (mS)	output admit. (mS)	junction cap. (fF)	gate–source cap. (fF)	gate–drain cap. (fF)
$g_{\mathrm{mp1}} = 95$	$g_{\mathrm{dsp1}} = 7$	$C_{\mathrm{jp1}} = 542$	$C_{\mathrm{gsp1}} = 305$	$C_{\mathrm{gdp1}} = 202$
$g_{\mathrm{mn1}} = 94$	$g_{\mathrm{dsn1}} = 6.4$	$C_{\mathrm{jn1}} = 21.1$	$C_{\mathrm{gsn1}} = 110$	$C_{\mathrm{gdn1}} = 47$
$g_{\mathrm{mp2}} = 45$	$g_{\mathrm{dsp2}} = 3.75$	$C_{\mathrm{jp2}} = 514$	$C_{\mathrm{gsp2}} = 140$	$C_{\mathrm{gdp2}} = 95$
$g_{\mathrm{mn2}} = 72$	$g_{\mathrm{dsn2}} = 4.43$	$C_{\mathrm{jn2}} = 20$	$C_{\mathrm{gsn2}} = 105$	$C_{\mathrm{gdn2}} = 35$
$g_{\mathrm{mp3}} = 18$	$g_{\mathrm{dsp3}} = 1.4$	$C_{\mathrm{jp3}} = 312$	$C_{\mathrm{gsp3}} = 60$	$C_{\mathrm{gdp3}} = 40$
$g_{\mathrm{mn3}} = 28$	$g_{\mathrm{dsn3}} = 1.7$	$C_{\mathrm{jn3}} = 7.8$	$C_{\mathrm{gsn3}} = 40$	$C_{\mathrm{gdn3}} = 28$

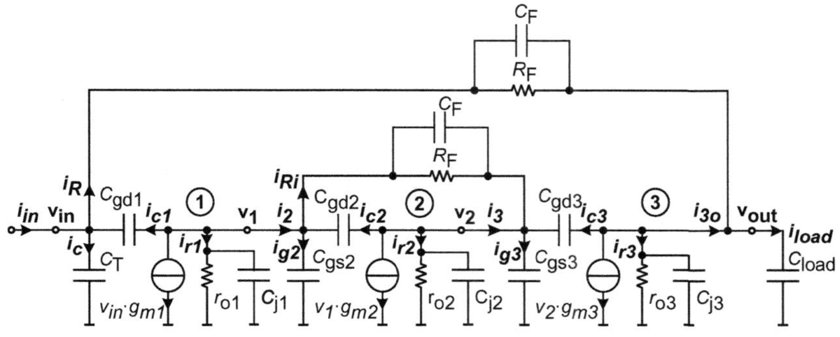

Fig. 6.31. Small-signal equivalent circuit of three-stage BM TIA with internal feedback (node numbers in circles)

C_{T} represents the capacitance of the external photodiode and the gate–source capacitances of the input transistors. The capacitance $C_{\mathrm{gd1}} = C_{\mathrm{gdn1}} + C_{\mathrm{gdp1}}$ also summarizes the gate–drain capacitances of both transistors N1 and P1, and the transconductance $g_{\mathrm{m1}} = g_{\mathrm{mn1}} + g_{\mathrm{mp1}}$ represents the sum of the transconductances of N1 and P1. For each amplifier stage the junction capacitances of n-MOSFET and p-MOSFET are combined to C_{j1} to C_{j3} at the nodes 1–3, respectively. P2 and P3 lead to the values of C_{gd2}, C_{gs2}, g_{m2}, C_{gd3}, C_{gs3}, and g_{m3}, respectively. The transistor loads N2 and N3 are represented as part of r_{o2} and r_{o3}.

Node equations are extracted from the small-signal equivalent circuit

$$i_{\mathrm{in}} = i_{\mathrm{R}} + i_{\mathrm{c}} - i_{\mathrm{c1}}, \tag{6.77}$$

$$-i_{\mathrm{c1}} = v_{\mathrm{in}} g_{\mathrm{m1}} + i_{\mathrm{r1}} + i_2, \tag{6.78}$$

$$i_2 = i_{\mathrm{g2}} + i_{\mathrm{Ri}} - i_{\mathrm{c2}}, \tag{6.79}$$

$$-i_{\mathrm{c2}} = v_1 g_{\mathrm{m2}} + i_{\mathrm{r2}} + i_3, \tag{6.80}$$

$$i_3 = i_{\mathrm{g3}} - i_{\mathrm{Ri}} - i_{\mathrm{c3}}, \tag{6.81}$$

$$-i_{\mathrm{c3}} = v_2 g_{\mathrm{m3}} + i_{\mathrm{r3}} + i_{\mathrm{3o}}, \tag{6.82}$$

$$i_{3o} = -i_R + i_{load}, \tag{6.83}$$

as well as the mesh equations presented as:

$$v_{in} = \frac{i_c}{j2\pi f C_T}, \tag{6.84}$$

$$v_{in} - v_1 = \frac{-i_{c1}}{j2\pi f C_{gd1}}, \tag{6.85}$$

$$v_1 = \frac{i_{r1}}{\frac{1}{r_{o1}} + j2\pi f C_{j1}}, \tag{6.86}$$

$$v_1 = \frac{i_{g2}}{j2\pi f C_{gs2}}, \tag{6.87}$$

$$v_1 - v_2 = \frac{-i_{c2}}{j2\pi f C_{gd2}}, \tag{6.88}$$

$$v_1 - v_2 = \frac{i_{Ri}}{\frac{1}{R_f} + j2\pi f C_f}, \tag{6.89}$$

$$v_2 = \frac{i_{r2}}{\frac{1}{r_{o2}} + j2\pi f C_{j2}}, \tag{6.90}$$

$$v_2 = \frac{i_{g3}}{j2\pi f C_{gs3}}, \tag{6.91}$$

$$v_2 - v_{out} = \frac{-i_{c3}}{j2\pi f C_{gd3}}, \tag{6.92}$$

$$v_{out} = \frac{i_{r3}}{\frac{1}{r_{o3}} + j2\pi f C_{j3}}, \tag{6.93}$$

$$v_{out} = \frac{i_{load}}{j2\pi f C_{load}}, \tag{6.94}$$

$$v_{in} - v_{out} = \frac{i_R}{\frac{1}{R_f} + j2\pi f C_f}. \tag{6.95}$$

Solving these equations with Maple for the parameters summarized in Table 6.7 leads to an effective low-frequency transimpedance of A_{IFB0} of $-8,628\,\Omega$, with a -3 dB bandwidth of 920 MHz.

Table 6.7. Transistor data belonging to Fig. 6.31

transcond. (mS)	output res. (Ω)	junction cap. (fF)	gate–source cap. (fF)	gate–drain cap. (fF)
$g_{m1} = 189$	$r_{o1} = 74.6$	$C_{j1} = 563$	$C_{gs1} = 414$	$C_{gd1} = 202$
$g_{m2} = 45$	$r_{o2} = 64.5$	$C_{j2} = 534$	$C_{gs2} = 140$	$C_{gd2} = 95$
$g_{m3} = 18.5$	$r_{o3} = 169.2$	$C_{j3} = 320$	$C_{gs3} = 60$	$C_{gd3} = 40$

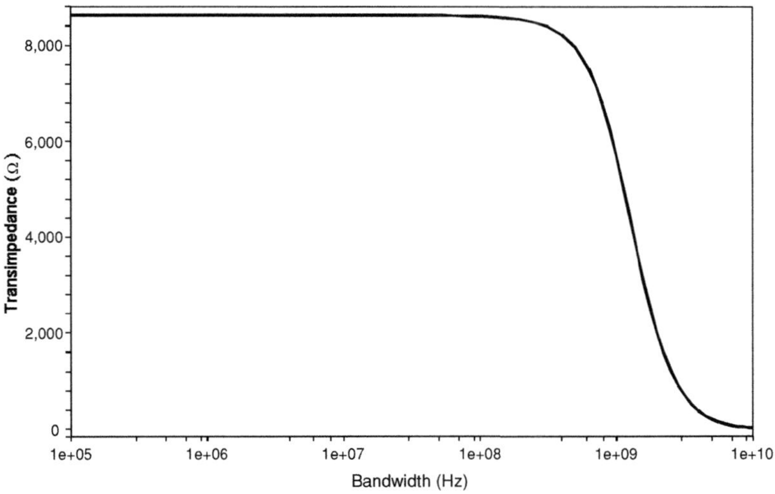

Fig. 6.32. Frequency response of simplified three-stage BM TIA with internal feedback with maximum feedback resistor $R_{\mathrm{F}} = 8{,}700\,\Omega$

6.3.5 Transimpedance-Gain Switching and Stability of Three-Stage Burst-Mode TIA with Internal Feedback

Similar to the stability analysis in Sect. 6.3.2 the root locus of the system is plotted. As described in (6.74) and (6.75) the characteristic polynomial in dependence on $s = \mathrm{j}2\pi f$ of can be derived by

$$P_{\mathrm{char}}(s) = 1 + \frac{A_{\mathrm{IFB}}(s)}{1 - A_{\mathrm{IFB}}(s)} \tag{6.96}$$

with $A_{\mathrm{IFB}}(s) = v_{\mathrm{out}}/i_{\mathrm{in}}$.

The actual circuit of the TIA with internal feedback is shown in Fig. 6.33. The feedback network consists of the transistors P4 to P6 and N5 and N6 and P4′ to P6′ and N5′ and N6′, respectively. This network is necessary to compensate the TIA in case of a small feedback resistance and keep the system stable. P4 and P5 as well as P4′ and P5′ form the actual feedback resistance depicted by R_{F} in the small equivalent circuit shown in Fig. 6.34. Furthermore, the transistors P6/P6′ are represented by their on-resistance R_{f3} and their parasitic capacitance C_{f3}. R_{f1} and C_{f1} stand for N4/N4′, and R_{f2} and C_{f2} for N5/N5′.

Again the effective transimpedance A_{IFB}, this time in dependence on $s = \mathrm{j}2\pi f$, is calculated. Therefore additional mesh and node equations are necessary for the feedback network:

$$i_{\mathrm{Rv1}} = i_{\mathrm{Rv2}} + i_{\mathrm{Rf}}, \tag{6.97}$$

$$i_{\mathrm{Rv1i}} = i_{\mathrm{Rv2i}} + i_{\mathrm{Rfi}}, \tag{6.98}$$

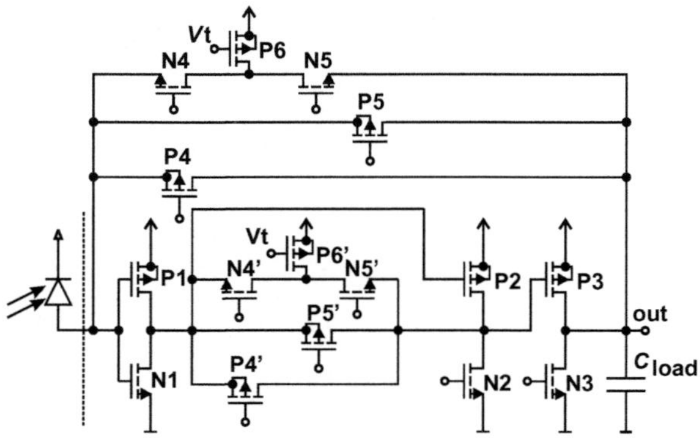

Fig. 6.33. Three-stage TIA with internal feedback with variable feedback network to guarantee stability

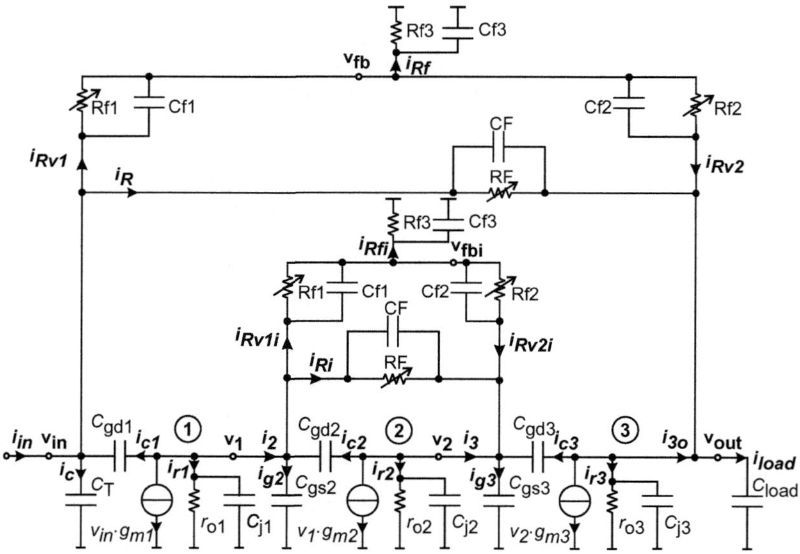

Fig. 6.34. Small-signal equivalent circuit of three-stage BM TIA with internal feedback concerning to Fig. 6.33

$$v_{\mathrm{fb}} = \frac{i_{\mathrm{Rf}}}{\frac{1}{R_{\mathrm{f3}}} + sC_{\mathrm{f3}}}, \tag{6.99}$$

$$v_{\mathrm{fbi}} = \frac{i_{\mathrm{Rfi}}}{\frac{1}{R_{\mathrm{f3}}} + sC_{\mathrm{f3}}}, \tag{6.100}$$

$$v_{\mathrm{in}} - v_{\mathrm{fb}} = \frac{i_{\mathrm{Rv1}}}{\frac{1}{R_{\mathrm{f1}}} + sC_{\mathrm{f1}}}, \tag{6.101}$$

$$v_{\mathrm{fb}} - v_{\mathrm{out}} = \frac{i_{\mathrm{Rv2}}}{\frac{1}{R_{\mathrm{f2}}} + sC_{\mathrm{f2}}}, \tag{6.102}$$

$$v_1 - v_{\text{fbi}} = \frac{i_{\text{Rv1i}}}{\frac{1}{R_{\text{f1}}} + sC_{\text{f1}}}, \tag{6.103}$$

$$v_{\text{fbi}} - v_2 = \frac{i_{\text{Rv2i}}}{\frac{1}{R_{\text{f2}}} + sC_{\text{f2}}}, \tag{6.104}$$

and (6.77) is substituted by

$$i_{\text{in}} = i_{\text{R}} + i_{\text{Rv1}} + i_C - i_{\text{C1}}, \tag{6.105}$$

(6.79) by

$$i_2 = i_{\text{g2}} + i_{\text{Ri}} + i_{\text{Rv1i}} - i_{\text{C2}}, \tag{6.106}$$

(6.81) by

$$i_3 = -i_{\text{Ri}} - i_{\text{Rv2i}} + i_{\text{g3}} - i_{\text{C3}}, \tag{6.107}$$

and (6.83) by

$$i_{3\text{o}} = -i_{\text{R}} - i_{\text{Rv2}} + i_{\text{load}}. \tag{6.108}$$

Figure 6.35 shows the root locus of the characteristic polynomial of the *three-stage BM TIA with internal feedback* for a maximum feedback resistor value $R_{\text{F}} = 8{,}700\,\Omega$. For the maximum feedback resistor value the open-loop gain of the amplifier has to be maximum and therefore the resistors R_{f1} and R_{f2} are switched off which is represented by very large values for these two resistors. The characteristic polynomial is an equation of the tenth order and is solved with Maple.

The values of the resistors and parasitic capacitances are summarized in Table 6.8. Due to the fact that all poles are situated in the left half-plane of the s-plane, the system is stable. (Please note the linear scaling of the axes. The dot closest to the y-axis is at the real position of -7.17×10^8.)

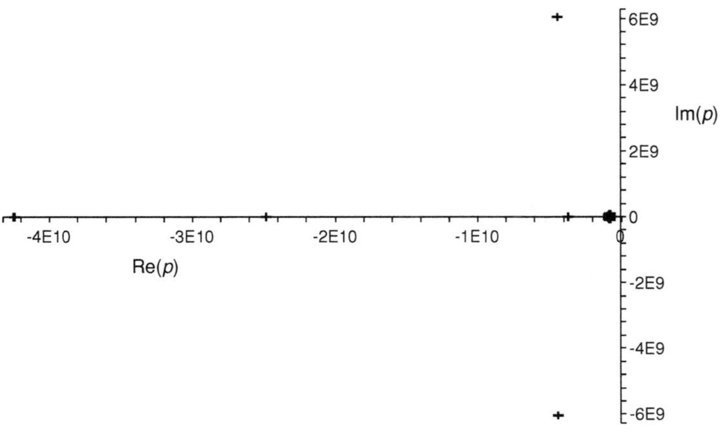

Fig. 6.35. Root locus of three-stage BM TIA with internal feedback with maximum feedback resistor R_{F}

Table 6.8. Values of the feedback network belonging to Fig. 6.33

resistances for max. feedback res. (Ω)	parasitic capacitance (fF)	resistance for min. feedback res. (Ω)
$R_{\mathrm{f}} = 8{,}700$	$C_{\mathrm{f}} = 5$	$R_{\mathrm{f}} = 141$
$R_{\mathrm{f1}} = 10^{10}$	$C_{\mathrm{f1}} = 40$	$R_{\mathrm{f1}} = 1{,}140$
$R_{\mathrm{f2}} = 10^{10}$	$C_{\mathrm{f2}} = 15$	$R_{\mathrm{f2}} = 600$
$R_{\mathrm{f3}} = 30$	$C_{\mathrm{f3}} = 80$	$R_{\mathrm{f3}} = 30$

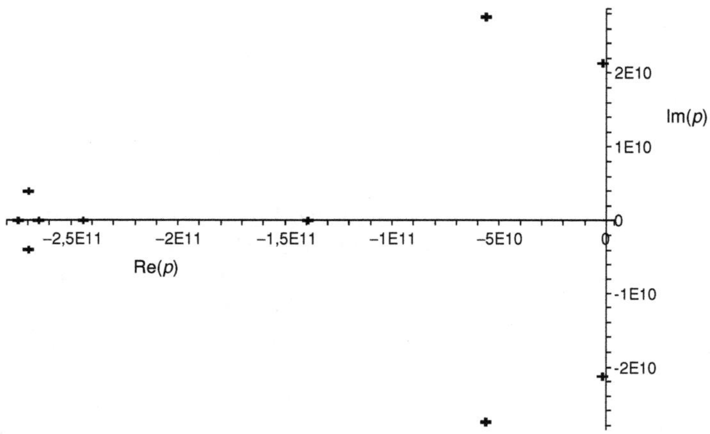

Fig. 6.36. Root locus of three-stage BM TIA with internal feedback for minimum feedback resistor R_{F}

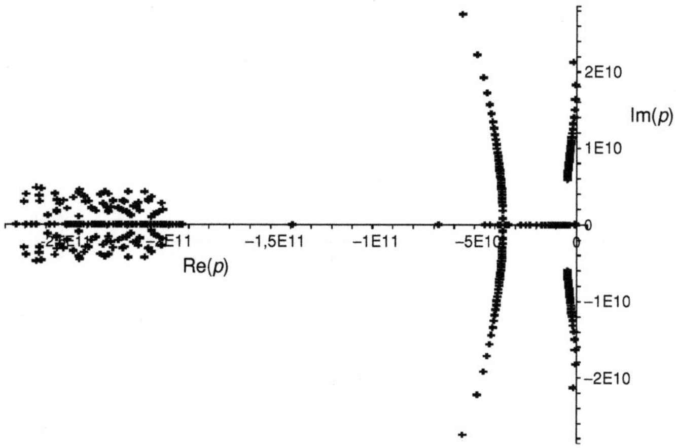

Fig. 6.37. Root locus of three-stage TIA with internal feedback with varied feedback network from minimum R_{F} to maximum R_{F}

For the case of the minimum feedback resistance the values of the resistors and capacitances are summarized in Table 6.8 also. This time the resistances R_{f1} and R_{f2} are switched on and therefore form an additional load for the amplifier stages, for a detailed description see Sect. 9.3.4. The root locus of the system with minimum feedback resistance is shown in Fig. 6.36. Again all poles are in the left half-plane (the two dots close to the y-axis are both at a real position of -1.6×10^9) and therefore the system is stable for this configuration.

Varying the feedback resistor requires also an adequate adjustment of the resistors R_{f1} and R_{f2}. Figure 6.37 shows the root locus for feedback resistors from minimum to maximum R_f and adequately adjusted R_{f1} and R_{f2}. Again all poles are moving in the left half-plane and therefore the system is stable for all feedback values between maximum and minimum R_F.

7

Noise Theory

Electronic noise has to be considered in the design of optical receivers. The received signal has to be stronger than the noise in order to avoid transmission errors. At the beginning of this chapter the important basics of sensitivity, bit-error rate (BER) and power penalty are discussed. All these factors and the electrical noise of the devices, which is discussed afterwards, influence the performance of the TIA. Noise models for TIAs with different kind of circuits are discussed, too.

7.1 Sensitivity and Power Penalty

The association between BER, sensitivity and power penalty are processed below.

7.1.1 Bit-Error Rate

Usually binary optical signals are used in optical data transmission leading to two discrete photocurrents $\langle i_0 \rangle$ and $\langle i_1 \rangle$ (Fig. 7.1). The expected value $\langle i_0 \rangle$ represents the logical zero and $\langle i_1 \rangle$ belongs to the logical one. It is allowed to assume that $\langle i_1 \rangle$ is larger than $\langle i_0 \rangle$. The photocurrent has a certain distribution around the expected or mean values due to current noise. The time slot T defines the bit rate B. For simplicity, an amplifier with ideal low-pass characteristics shall be assumed. The bandwidth of this amplifier will be assumed as $B/2$ [101]. The following derivation, however, can easily be done also for a bandwidth of $2B/3$ or $3B/4$ often used in practice.

The variance of the input noise current δi_{ges} is obtained by integration of the spectral noise density [101]:

$$\langle \delta i_{\mathrm{ges}}^2 \rangle = \int_0^{B/2} \langle \delta i^2(f) \rangle \mathrm{d}f \ , \tag{7.1}$$

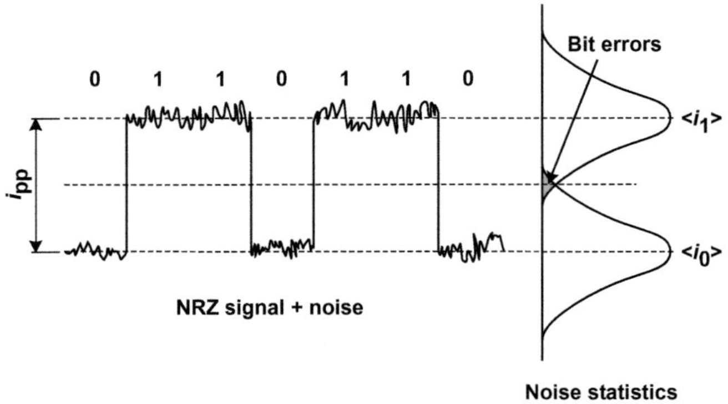

Fig. 7.1. Noisy non-return-to-zero input signal with noise statistics

where $g_{\mathrm{m}}R_{\mathrm{L}} \gg 1$ has been used. We assume that the instant current values have a Gaussian distribution around the expected values $\langle i_0 \rangle$ and $\langle i_1 \rangle$. The probability density for a logical zero, therefore, is

$$p_0(i) = \frac{1}{\sqrt{2\pi \langle \delta i_{\mathrm{ges}}^2 \rangle}} \exp\left(-\frac{(i - \langle i_0 \rangle)^2}{2\langle \delta i_{\mathrm{ges}}^2 \rangle} \right) . \tag{7.2}$$

For a logical one,

$$p_1(i) = \frac{1}{\sqrt{2\pi \langle \delta i_{\mathrm{ges}}^2 \rangle}} \exp\left(-\frac{(i - \langle i_1 \rangle)^2}{2\langle \delta i_{\mathrm{ges}}^2 \rangle} \right) \tag{7.3}$$

is obtained correspondingly. For equally distributed logical zeros and ones, the mean value of the current in respect to time is:

$$D_{\mathrm{t}} = (\langle i_1 \rangle + \langle i_0 \rangle)/2 . \tag{7.4}$$

This value is defined as the decision threshold. For a current $i \le D_{\mathrm{t}}$, the detected signal is considered as logical zero. The probability of a wrong decision is given by the BER for the logical zero

$$\mathrm{BER}_0 = \int_{D_{\mathrm{t}}}^{\infty} p_0(i)\mathrm{d}i = \frac{1}{\sqrt{2\pi}} \int_{Q}^{\infty} \exp\left(-u^2/2 \right) \mathrm{d}u , \tag{7.5}$$

with $u = (i - \langle i_0 \rangle)/\sqrt{\langle \delta i_{\mathrm{ges}}^2 \rangle}$ and where the noise distance,

$$Q = \frac{\langle i_1 \rangle - \langle i_0 \rangle}{2\sqrt{\langle \delta i_{\mathrm{ges}}^2 \rangle}} , \tag{7.6}$$

also has been introduced. For the applied assumptions it can be shown that the BER for the logical one is equal to that for the logical zero

$$\mathrm{BER}_1 = \int_{-\infty}^{D_{\mathrm{rmt}}} p_1(i)\mathrm{d}i = \mathrm{BER}_0 = \mathrm{BER}. \tag{7.7}$$

Therefore the BER in general obeys

$$\mathrm{BER} = \frac{1}{\sqrt{2\pi}} \int_Q^{\infty} \exp\left(-u^2/2\right) \approx \frac{1}{\sqrt{2\pi}} \exp\left(-\frac{Q^2}{2}\right)\left(1 - \frac{1}{Q^2}\right). \tag{7.8}$$

The error made by the approximation is less than 1% for $Q \geq 2$. The relation between BER and the noise distance Q is shown in Fig. 7.2. For instance, BER $= 10^{-9}$ holds at $Q = 6$ and BER$= 1.3 \times 10^{-12}$ for $Q = 7$.

Modern so-called communication analyzers or digital sampling oscilloscopes can be used to determine the noise distance Q. For this purpose, eye diagrams are taken with a set-up shown in Fig. 7.3.

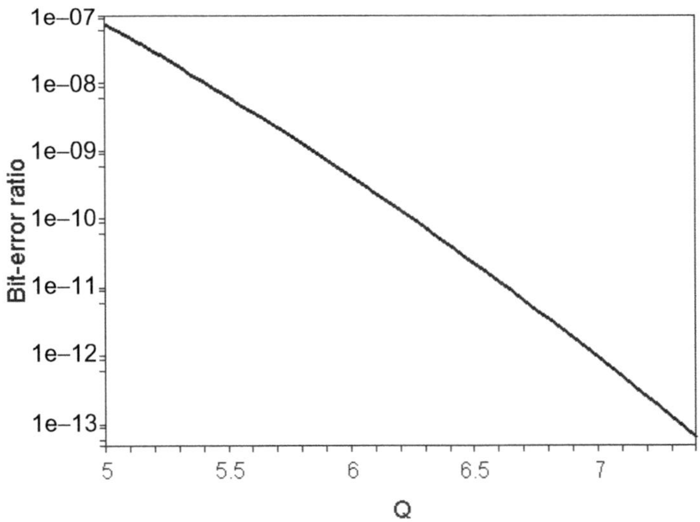

Fig. 7.2. BER over Q-factor [101]

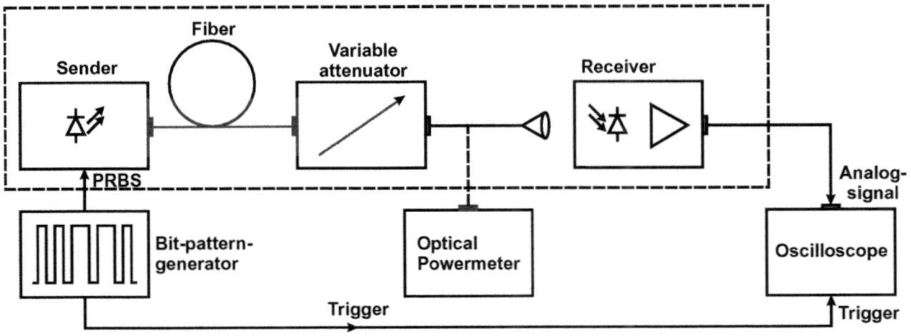

Fig. 7.3. Measurement set-up for eye diagram measurements

The receiver under test in Fig. 7.3 has an analog output capable of driving the 50 Ω input of the digital sampling oscilloscope. The digital sampling oscilloscope or communication analyzer determines the histograms shown in Fig. 7.4. The mean values $\langle i_0 \rangle$ and $\langle i_1 \rangle$ as well as their variances can be read from the display of the communication analyzer. From the variances, $\langle \delta i^2_{\mathrm{ges}} \rangle$ and finally the noise distance Q is calculated by the communication analyzer.

When the receiver under test has a digital output the set-up shown in Fig. 7.5 can be used for a bit-error analysis. A bit-error analyzer also can be used when the analog output of an optical receiver has a larger output voltage than the sensitivity of the bit-error analyzer. The bit-error analyzer digitally compares the received bits with the sent bits and counts errors. A bit-error analyzer results in a more accurate characterization of optical receivers than the determination of Q with the set-up of Fig. 7.3.

The decision threshold D_{t} is in the middle between the "1"- and "0"-photocurrents. For bit sequences with equally distributed "1"s and "0"s it, therefore, can be expressed by the mean optical power $\eta \langle P \rangle$, which is converted into photocurrent:

$$D_{\mathrm{t}} = \frac{\langle i_1 \rangle + \langle i_0 \rangle}{2} = \eta \frac{q}{\hbar\omega} \langle P \rangle. \tag{7.9}$$

The ratio r of the two photocurrent levels can be defined:

$$r = \langle i_0 \rangle / \langle i_1 \rangle. \tag{7.10}$$

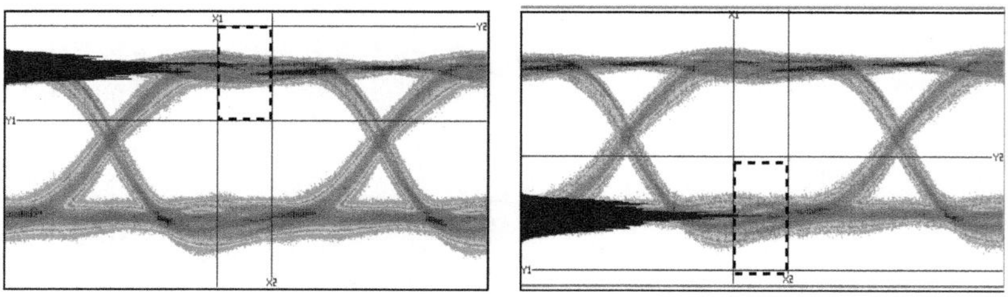

Fig. 7.4. Eye diagrams with histograms for the "1" (*left*) and "0" (*right*)

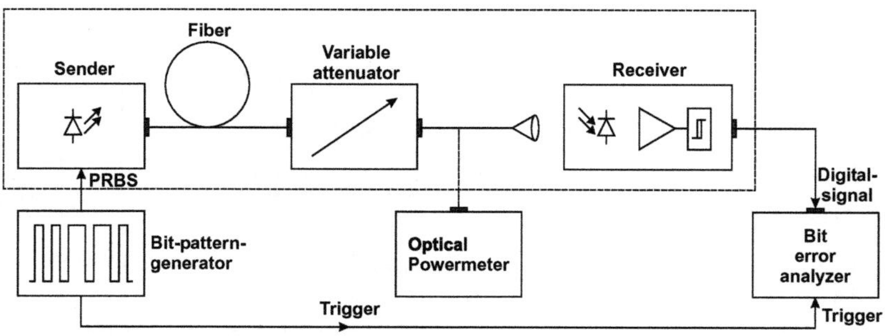

Fig. 7.5. Measurement set-up for digital bit-error analysis

With this definition, the optical power of a binary signal necessary to achieve a certain required BER (i.e., the corresponding Q-factor) results:

$$\langle P \rangle = \frac{1+r}{1-r} \frac{\hbar\omega}{\eta q} Q \sqrt{\langle \delta i_{\text{ges}}^2 \rangle}. \tag{7.11}$$

This quantity is called the sensitivity of the optical receiver and it is usually expressed in dBm ($10 \times \log(\langle P \rangle / 1\,\text{mW})$). It becomes lowest for $r = 0$, i.e., $\langle i_0 \rangle = 0$ or a vanishing "0" optical power. In this case it is

$$\langle P \rangle = \frac{\hbar\omega}{\eta q} Q \sqrt{\langle \delta i_{\text{ges}}^2 \rangle}. \tag{7.12}$$

7.1.2 Sensitivity

The available optical input power depends on the optical input power of the transmitter and on the attenuation in the fiber, i.e., on the length of the fiber link. For long-distance optical transmission the sensitivity of the receiver must be high to compensate for the transmittance losses.

Together with the optical input power P_{opt} the responsivity R of the photodiode defines the input current I_{in} of the TIA.

$$I_{\text{in}} = P_{\text{opt}} R. \tag{7.13}$$

The responsivity of photodiodes depends on the wavelength and the external quantum efficiency of the photodiode (see Sect. 3). Therefore R of a photodiode for infrared light is higher than R of a photodiode for red light for the same external quantum efficiency. This leads directly to a better sensitivity for infrared light. The sensitivity difference, related to $\lambda = 1{,}540$ nm, is shown in Table 7.1.

7.1.3 Power Penalty

Power penalty defines the loss of sensitivity due to the fact that the photocurrent value for logic zero is bigger than zero. In this case the opening of the eye

Table 7.1. Sensitivity difference for the same photodiode quantum efficiency due to responsivity differences for different wavelengths, related to λ =1540 nm

wavelength (nm)	Δ sensitivity (dB)
1,540	0
1,300	0.75
650	3.75

is smaller than expected for ideal $i_0 = 0$. The specification of the laser for the burst-mode application defines a "0"-signal of a tenth of the optical power of the "1"-signal. This leads to an extinction ratio of 10.

For lasers providing a finite extinction ratio (EX) the measured sensitivity has to be corrected by the power penalty due to the extinction ratio. For example if the extinction ratio of the laser equals three, the eye opening of two-thirds of the maximum optical input power can be converted to the opening corresponding to a laser source with an extinction ratio of 10, what comes up for burst-mode applications. Figure 7.6 shows the principle. Only the average optical input power $\overline{P}_{\text{opt}}$ can be measured, since the optical power meter used to determine the optical input power of the receiver is too slow for peak power detection [102]. The correction factor for a laser source with EX = 3–10 will be calculated.

We obtain

$$\overline{P}_{\text{opt}} = \frac{P_{\text{opt,max}} - P_{\text{opt,min}}}{2} + P_{\text{opt,min}}, \tag{7.14}$$

where $P_{\text{opt,min}}$ is the optical input power, for logic "0", and $P_{\text{opt,max}}$ is the optical input power for logic "1". Equation (7.14) leads to (7.15) for the given example with EX = 3.

$$\overline{P}_{\text{opt}} = \frac{P_{\text{opt,max}} - \frac{1}{3}P_{\text{opt,max}}}{2} + \frac{1}{3}P_{\text{opt,max}} = \frac{2}{3}P_{\text{opt,max}}. \tag{7.15}$$

The average optical input power can also be calculated in dependence on $P_{\text{opt,max,corr}}$ for EX = 10, see (7.16).

$$\overline{P}_{\text{opt,corr}} = \frac{P_{\text{opt,max,corr}} - P_{\text{opt,min,corr}}}{2} + P_{\text{opt,min,corr}} = 0.55P_{\text{opt,max,corr}}. \tag{7.16}$$

The optical input power swing ΔP_{opt} and $\Delta P_{\text{opt,corr}}$, respectively, for both extinction ratios must be the same, therefore we assume

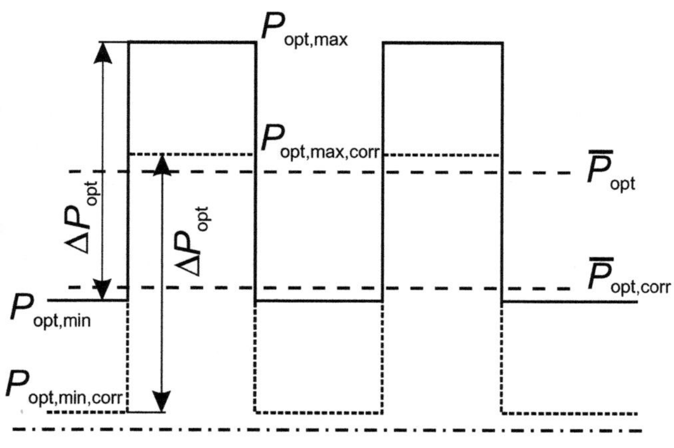

Fig. 7.6. Power penalty due to extinction ratio of the laser

$$\Delta P_{\mathrm{opt}} = \Delta P_{\mathrm{opt,corr}}. \tag{7.17}$$

Substituting (7.15) and (7.16) leads to

$$P_{\mathrm{opt,max,corr}} = \frac{\frac{2}{3} P_{\mathrm{opt,max}}}{0.9} = 0.55 P_{\mathrm{opt,max}}, \tag{7.18}$$

which leads to

$$\overline{P}_{\mathrm{opt,corr}} = 0.407 P_{\mathrm{opt,max}} = 0.61 \overline{P}_{\mathrm{opt}}. \tag{7.19}$$

Usually the average optical input power is given in dBm and therefore the correction factor in dB is -2.1 dB for measurements with EX = 3 and required sensitivities with EX = 10:

$$\overline{P}_{\mathrm{opt,corr}}[\mathrm{dBm}] = \overline{P}_{\mathrm{opt}}[\mathrm{dBm}] - 2.1\,\mathrm{dBm}. \tag{7.20}$$

7.2 Noise Models of Components

This section summarizes the noise models of the main noise sources of common amplifiers. Models for resistor noise, bipolar, and heterojunction bipolar transistor noise and an MOSFET noise model are presented. Before the noise models of the devices are discussed, the main sources of noise will be summarized.

The *shot noise* is always coupled to a direct current flow and is present in diodes and bipolar transistors. The external current through a diode or a bipolar transistor seems to be a steady flow of current, but it is in fact the sum of different independent current pulses with an average current I_{av}. The fluctuation I is termed *shot noise* and is generally specified in terms of its mean-square variation about the average value [103]. This value $\overline{i^2}$ is calculated as follows:

$$\begin{aligned} \overline{i^2} &= \overline{(I - I_{\mathrm{av}})^2} \\ &= \lim_{T_{\mathrm{p}} \to \infty} \frac{1}{T_{\mathrm{p}}} \int_0^{T_{\mathrm{p}}} (I - I_{\mathrm{av}})^2 \, \mathrm{d}t \end{aligned} \tag{7.21}$$

where T_{p} equals the period interval, which approaches infinity. As mentioned before the current I consists of random independent pulses with the average value I_{av}. For this case it can be shown that the resulting noise current has the average value $\overline{i^2}$

$$\overline{i^2} = 2 q I_{\mathrm{av}} \Delta f \tag{7.22}$$

where q is the electronic charge. This value can be described as a constant function of the frequency, the noise current density $\overline{i^2}/\Delta f$ with the unit of $\mathrm{A}^2\,\mathrm{Hz}^{-1}$. Therefore, the noise current is directly proportional to the square root of the measurement bandwidth.

Thermal noise is generated due to the random thermal electrons in conventional resistors. It is unaffected by an existing or nonexisting direct current since the drift velocities in a conductor are much lower than thermal velocities of electrons. Therefore it is directly dependent on the temperature T. Thermal noise is only zero if the temperature drops to absolute zero.

Another type of noise found in every active device is *flicker noise*. There are different reasons for flicker noise, but in bipolar transistors it is caused mainly by traps associated with contamination and crystal defects in the emitter-base depletion layer. These traps catch and release electrons as a random process and the time constants associated with this process give rise to a noise signal with high noise levels at low frequencies [103]. Flicker noise is usually much stronger in MOSFETs due to Si/SiO_2-interface states and current flows just along this interface.

Flicker noise is always related to direct current flow and shows a spectral density dependence on $1/f$, see (7.23). This $1/f$ dependence is the reason for the alternative name $1/f$ *noise*

$$\overline{i^2} = K_1 \frac{I^a}{f^b} \Delta f. \tag{7.23}$$

I is a direct current, K_1 is a constant for a particular device, a is a constant between 0.5 and 2 and b is a constant around unity. The values of the constants show a big difference for different devices and technologies. Usually they are found empirically.

Other existing noise sources, for example burst noise and avalanche noise, are not discussed, because they do not influence the following calculations. They can be looked up in the literature, for example, in [103].

7.2.1 Resistor Noise Model

Monolithic and thin-film resistors show thermal noise. The spectral noise voltage density $\overline{v_{\mathrm{R}}^2}$ is given by

$$\overline{v_{\mathrm{R}}^2} = 4k_{\mathrm{B}}TR\Delta f, \tag{7.24}$$

where k_{B} is the Boltzmann constant and T the temperature. Equation (7.25) displays the spectral noise current density $\overline{i_{\mathrm{R}}^2}$

$$\overline{i_{\mathrm{R}}^2} = 4k_{\mathrm{B}}T\frac{1}{R}\Delta f. \tag{7.25}$$

Equations (7.24) and (7.25) show that the spectral noise densities are independent of the frequency and therefore thermal noise is a source of white noise. The spectral noise current density can be calculated from the spectral noise voltage density as

$$\overline{i_{\mathrm{R}}^2} = \frac{\overline{v_{\mathrm{R}}^2}}{R^2}. \tag{7.26}$$

Figure 7.7a shows the equivalent circuit of the resistor with noise current source and Fig. 7.7b shows the resistor with the noise voltage source.

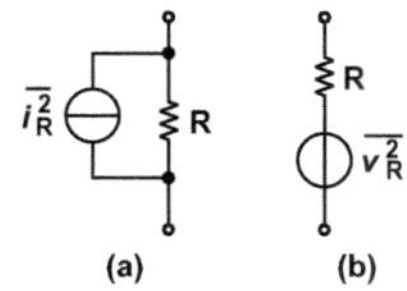

Fig. 7.7. (**a**) Resistor noise current source and (**b**) resistor noise voltage source

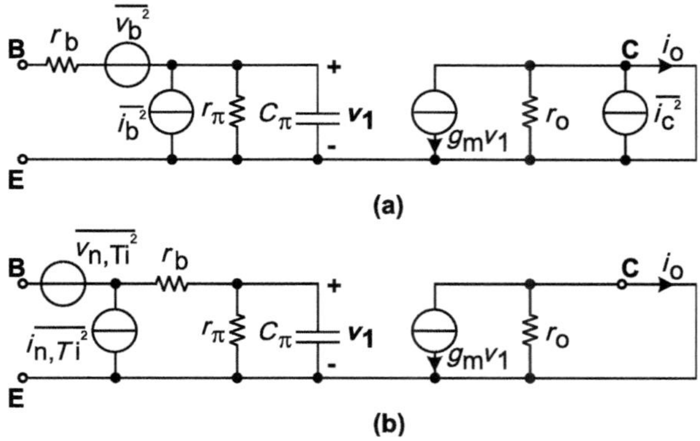

Fig. 7.8. (**a**) Noise sources in bipolar transistors and (**b**) equivalent transistor input noise sources

7.2.2 Bipolar- and Heterojunction-Bipolar-Transistor Noise Model

In a bipolar transistor in the forward-active region, minority carriers diffuse and drift across the base region to be collected at the collector–base junction. Minority carriers entering the collector–base depletion region are accelerated by the field existing there and swept across this region to the collector. The time of arrival at the collector–base junction of the diffusing (or drifting) carriers is a purely random process, and thus the transistor collector current consists of a series of random current pulses. Consequently, the collector current I_C shows full shot noise as given by (7.22), and this is represented by a shot noise current generator $\overline{i_c^2}$ from collector to emitter as shown in the equivalent circuit in Fig. 7.8a [103].

The base current I_B in a transistor is due to recombination in the base and base–emitter depletion regions and also to carrier injection from the base into the emitter. All of theses are independent random processes, and thus I_B also shows full shot noise. This is represented by the shot noise current generator $\overline{i_b^2}$ in Fig. 7.8a. The collector series resistor r_c also shows thermal noise, but since this resistor is in series with the high-impedance collector node, this noise is negligible and is usually not included into the model. The resistors r_π and r_o in Fig. 7.8a are not real resistors; they are for modeling purposes only and therefore do not show thermal noise. The so-called intrinsic base region below the emitter is connected laterally by the so-called extrinsic base this leads to a series resistor. Because of a minimum required distance of extrinsic base from the emitter and the rather high sheet resistance of the base a series resistance r_b exists in the base of bipolar transistors (see Fig. 7.9). The base resistor r_b is a physical resistor and thus adds thermal noise. This noise is represented in the noise source $\overline{v_\mathrm{b}^2}$,

$$\overline{v_\mathrm{b}^2} = 4k_\mathrm{B}Tr_\mathrm{b}\Delta f. \tag{7.27}$$

The collector current shot noise is given by (7.28). Flicker noise and burst noise were experimentally found in bipolar transistors and represented as noise generators across the internal base–emitter junction. These two noise generators and the base current shot noise are combined to the current noise source $\overline{i_\mathrm{b}^2}$ shown in (7.29). Usually, however, Flicker noise and burst noise are negligible in bipolar transistors

$$\overline{i_\mathrm{c}^2} = 2qI_\mathrm{C}\Delta f, \tag{7.28}$$

$$\overline{i_\mathrm{b}^2} = \underbrace{2qI_\mathrm{B}\Delta f}_{\text{Shot noise}} + \underbrace{K_1\frac{I_\mathrm{B}^a}{f}\Delta f}_{\text{Flicker noise}} + \underbrace{K_2\frac{I_\mathrm{B}^c}{1+(f/f_\mathrm{c})^2}\Delta f}_{\text{Burst noise}}. \tag{7.29}$$

The noise sources defined in (7.27)–(7.29) can be represented as input noise sources $\overline{v_\mathrm{n,Ti}^2}$ and $\overline{i_\mathrm{n,Ti}^2}$ shown in Fig. 7.8b. The output noise current i_o of the circuit in Fig. 7.8a is calculated with short circuited input-node and output-node. The equivalent input noise voltage $\overline{v_\mathrm{n,Ti}^2}$ of Fig. 7.8b has to show the same output noise as the circuit with original noise sources:

$$g_\mathrm{m}v_\mathrm{b} + i_\mathrm{c} = g_\mathrm{m}v_\mathrm{n,Ti}. \tag{7.30}$$

With short circuited input, $\overline{i_\mathrm{b}^2}$ could be neglected due to $r_\mathrm{b} \ll r_\pi$. We obtain for $v_\mathrm{n,Ti}$ from (7.30):

$$v_\mathrm{n,Ti} = v_\mathrm{b} + \frac{i_\mathrm{c}}{g_\mathrm{m}} \tag{7.31}$$

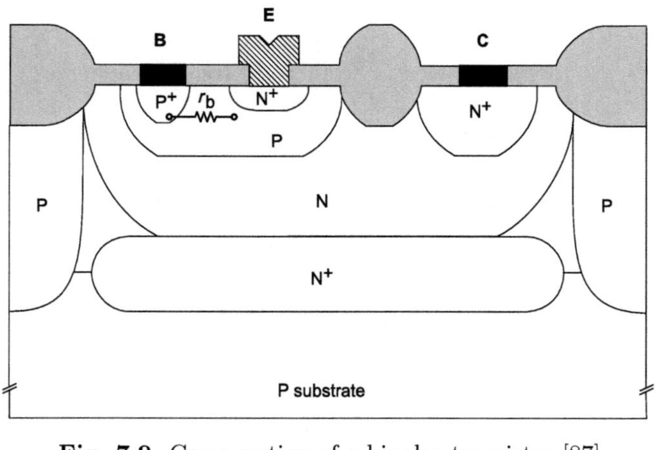

Fig. 7.9. Cross section of a bipolar transistor [87]

due to independent i_c and v_b, (7.31) leads to

$$\overline{v_{\mathrm{n,Ti}}^2} = \overline{v_\mathrm{b}^2} + \frac{\overline{i_\mathrm{c}^2}}{g_\mathrm{m}^2}; \tag{7.32}$$

substituting the noise sources $\overline{v_\mathrm{b}^2}$ and $\overline{i_\mathrm{c}^2}$ calculated in (7.27) and (7.28):

$$\overline{v_{\mathrm{n,Ti}}^2} = 4k_\mathrm{B}Tr_\mathrm{b}\Delta f + \frac{2qI_\mathrm{C}\Delta f}{g_\mathrm{m}^2}. \tag{7.33}$$

The basic equations for I_C and g_m of the bipolar transistor are shown in (7.34) and (7.35):

$$I_\mathrm{C} = I_\mathrm{S}\left(1 + \frac{V_\mathrm{CE}}{V_\mathrm{Ea}}\right)\exp\left(\frac{V_\mathrm{BE}}{\frac{k_\mathrm{B}T}{q}}\right), \tag{7.34}$$

$$g_\mathrm{m} = \frac{\partial I_\mathrm{C}}{\partial V_\mathrm{BE}} = \frac{I_\mathrm{C}}{\frac{k_\mathrm{B}T}{q}} = \frac{I_\mathrm{C}}{V_\mathrm{T}}. \tag{7.35}$$

Equation (7.33) leads with (7.34) and (7.35) [13] to:

$$\frac{\overline{v_{\mathrm{n,Ti}}^2}}{\Delta f} = 4k_\mathrm{B}T\left(r_\mathrm{b} + \frac{1}{2g_\mathrm{m}}\right). \tag{7.36}$$

For the calculation of the equivalent input noise current $\overline{i_{\mathrm{n,Ti}}^2}$ of the transistor, the inputs of both circuits shown in Fig. 7.8 are open circuited and the load is still short circuited. Again the output noise currents i_0 due to the original sources and the equivalent input noise current, respectively, are equated. With rms noise quantities we calculate as follows:

$$\beta(\mathrm{j}\omega)i_{\mathrm{n,Ti}} = i_\mathrm{c} + \beta(\mathrm{j}\omega)i_\mathrm{b}. \tag{7.37}$$

This gives

$$i_{\mathrm{n,Ti}} = i_\mathrm{b} + \frac{i_\mathrm{c}}{\beta(\mathrm{j}\omega)} \tag{7.38}$$

and since i_b and i_c are independent generators (7.38) leads to

$$\overline{i_{\mathrm{n,Ti}}^2} = \overline{i_\mathrm{b}^2} + \frac{\overline{i_\mathrm{c}}}{|\beta(\mathrm{j}\omega)|^2} \tag{7.39}$$

with $\omega = 2\pi f$ and

$$\beta(\mathrm{j}\omega) = \frac{\beta_0}{1 + \mathrm{j}\left(\omega/\omega_\beta\right)}, \tag{7.40}$$

where β_0 is the low-frequency, small-signal current gain of the bipolar transistor. Substituting (7.29), (7.28) and (7.40) into (7.39) leads to

$$\frac{\overline{i_{\mathrm{n,Ti}}^2}}{\Delta f} = 2q\left(I_\mathrm{B} + \frac{K_1}{2q}\frac{I_\mathrm{B}^a}{f} + \frac{I_\mathrm{C}}{|\beta(\mathrm{j}f)|^2}\right). \tag{7.41}$$

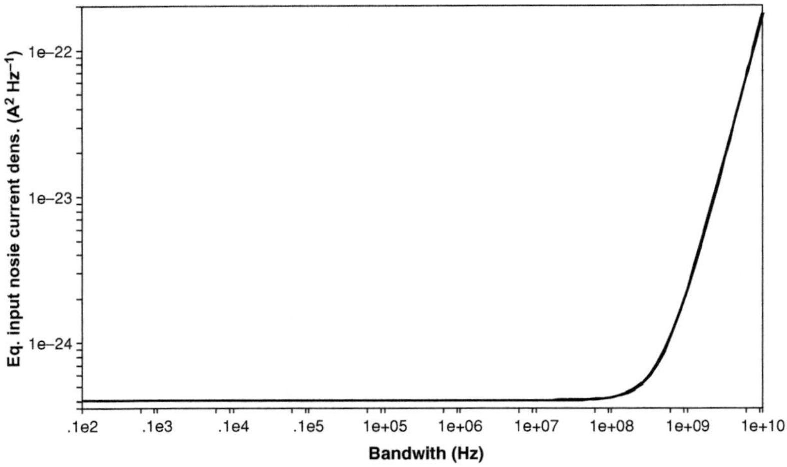

Fig. 7.10. Equivalent input noise current density of bipolar transistor, $I_C = 100\,\mu A$, $\beta = 80$, $f_\beta = 500\,MHz$, flicker noise neglected

Figure 7.10 displays the equivalent input noise current density of a bipolar transistor with $I_C = 100\,\mu A$, $\beta = 80$, and $f_\beta = 500\,MHz$, with f_β being the $-3\,dB$ cut-off frequency of the current gain β of the bipolar transistor. The equivalent input noise current density at low frequencies is dominated by flicker noise and although this portion looks quite large it can be neglected for high bandwidths of more than several hundred megahertz, because the equivalent input noise density is increasing with the squared frequency, which is the main noise contribution for high frequencies (see Fig. 7.10) especially for high input-node capacitances.

The major differences between Si bipolar transistors (BJT) and SiGe heterojunction bipolar transistors (HBT) are:

1. The base of the SiGe HBT is higher doped leading to a lower r_b
2. β of the SiGe HBT is larger
3. The transit frequency f_T of the HBT is larger due to the gradient in the electric field in the base coming from the Ge gradient in the base

For the same bandwidth of a circuit, the circuit with SiGe HBT,therefore, shows less noise than that with an Si BJT.

7.2.3 Field-Effect-Transistor Noise Model

To evaluate the variance of the input noise current $\langle \delta i_{ges}^2 \rangle$ the different spectral noise densities are calculated below. Cross sections of MOSFETs can be found in the literature, for example, in [87, 104].

The resistive channel of field-effect transistors (FETs) joining source and drain is modulated by the gate–source voltage so that the drain current is

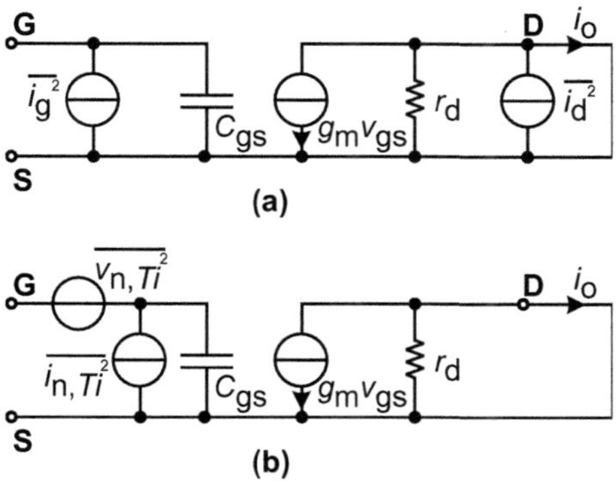

Fig. 7.11. Small-signal equivalent circuit for FETs: (**a**) noise sources in FETs; (**b**) equivalent transistor input noise sources

controlled by the gate–source voltage. Since the channel material is resistive it exhibits thermal noise, this is the main noise source in FETs. It can be shown that this noise source can be represented by a noise–current generator $\overline{i_\mathrm{d}^2}$ (see Fig. 7.11a) from drain to source in the small-signal equivalent circuit. Flicker noise, which is found experimentally in the FET, is represented by a drain–source current generator. These two are combined into one noise source (see (7.42)) [103]

$$\overline{i_\mathrm{d}^2} = \underbrace{4k_\mathrm{B}T\Gamma_\mathrm{F}g_\mathrm{m}\Delta f}_{\text{thermal noise}} + \underbrace{K\frac{I_\mathrm{D}^a}{f}\Delta f}_{\text{flicker noise}} \; . \tag{7.42}$$

The other noise source is the shot noise $\overline{i_\mathrm{g}^2}$ generated by the gate leakage current and is therefore modeled as a noise source between gate and source as you can see in Fig. 7.11a. Due to very small gate currents of about several pA in MOSFETs (as long as no tunneling currents are involved as will happen for 65 nm CMOS!), $\overline{i_\mathrm{g}^2}$ is usually very small and can be neglected for most cases

$$\overline{i_\mathrm{g}^2} = 2qI_\mathrm{G}\Delta f. \tag{7.43}$$

The equivalent input noise sources can be calculated out of the small-signal equivalent circuit. Figure 7.11b shows the equivalent input noise sources. For calculating the equivalent input voltage at first, the output noise is calculated with short-circuited load. The output noise current i_0 for the original noise sources and for the equivalent input noise voltage has to be equal [103]. The equivalent input noise voltage source $\overline{v_\mathrm{n,Ti}^2}$ has to generate the output noise current given in (7.42). The equivalent input noise voltage $\overline{v_\mathrm{n,Ti}^2}$ is amplified according to the transconductance g_m.

$$i_\mathrm{d} = g_\mathrm{m}v_\mathrm{n,Ti} \tag{7.44}$$

which leads to

$$\overline{v_{\mathrm{n,Ti}}^2} = \frac{\overline{i_{\mathrm{d}}^2}}{g_{\mathrm{m}}^2}. \tag{7.45}$$

With (7.42) and neglecting (7.43) we obtain

$$\frac{\overline{v_{\mathrm{n,Ti}}^2}}{\Delta f} = 4k_{\mathrm{B}}T\frac{2}{3}\frac{1}{g_{\mathrm{m}}} + K\frac{I_{\mathrm{D}}^a}{g_{\mathrm{m}}^2 f}. \tag{7.46}$$

The equivalent input noise current is calculated with open input node and again short-circuited load. This leads to

$$i_{\mathrm{n,Ti}}\frac{g_{\mathrm{m}}}{\mathrm{j}\omega C_{\mathrm{gs}}} = i_{\mathrm{d}} \tag{7.47}$$

and again with independent noise sources, (7.47) leads to (7.48).

$$\overline{i_{\mathrm{n,Ti}}^2} = \overline{i_{\mathrm{g}}^2} + \overline{i_{\mathrm{d}}^2}\frac{\omega^2 C_{\mathrm{gs}}^2}{g_{\mathrm{m}}^2}. \tag{7.48}$$

Neglecting i_{g}, this gives with (7.42) the equivalent input noise current density $\overline{i_{\mathrm{n,Ti}}^2}$:

$$\frac{\overline{i_{\mathrm{n,Ti}}^2}}{\Delta f} = \omega^2 C_{\mathrm{gs}}^2 \left(4k_{\mathrm{B}}T\frac{2}{3}\frac{1}{g_{\mathrm{m}}} + K\frac{I_{\mathrm{D}}^a}{g_{\mathrm{m}}^2 f}\right). \tag{7.49}$$

Figure 7.12 shows the equivalent input noise current density over the bandwidth with $I_{\mathrm{D}} = 100\,\mu\mathrm{A}$, $C_{\mathrm{gs}} = 0.5\,\mathrm{pF}$, and $g_{\mathrm{m}} = 10\,\mathrm{mS}$. Flicker noise is neglected, due to the fact that for high bandwidths the amount of flicker noise is small compared to noise current at high frequencies, where the noise density is increasing with the squared frequency.

If we compare Figs. 7.12 and 7.10, we see that the MOSFET has a larger input noise current density at 1 GHz than the bipolar transistor.

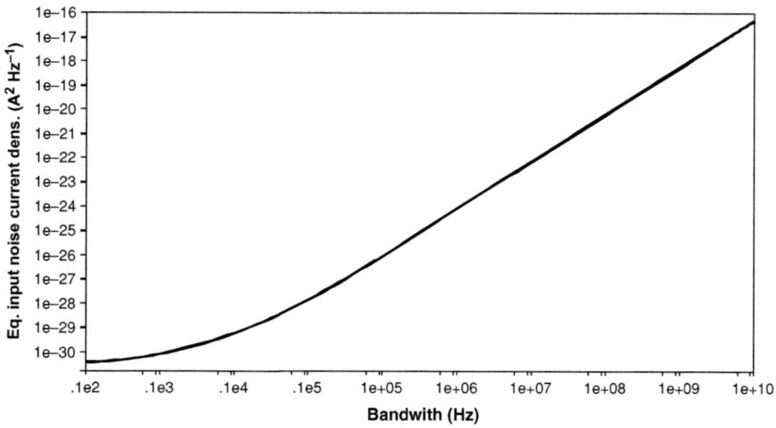

Fig. 7.12. Equivalent input noise current density of FET over the bandwidth, $I_{\mathrm{D}} = 100\,\mu\mathrm{A}$, $C_{\mathrm{gs}} = 0.5\,\mathrm{pF}$, $g_{\mathrm{m}} = 10\,\mathrm{mS}$, flicker noise neglected

7.3 Noise Models of Transimpedance Amplifier

In the following chapter noise models of transimpedance amplifiers (TIAs) will be presented. First a TIA with an ideal amplifier will be discussed. Two different TIAs with MOSFET input-stages are described in detail subsequently.

7.3.1 Ideal-Amplifier TIA

First the general model of the ideal TIA is considered. In Fig. 7.13 the noise sources of the individual TIA components are included. The noise of the TIA depends on the technology and topology of the circuit.

As a first estimation we define a very simple circuit which also converts an input current into an output voltage, a simple pin photodiode with following amplifier [101]. In the easiest case this amplifier is a common emitter or a common drain circuit, respectively. With more complicated designs better results can be achieved, but as an upper bound for noise we consider the circuit shown in Fig. 7.14. The main noise sources are the thermal noise of the resistor R_{F} and the input noise sources of the amplifier $\overline{i^2_{\mathrm{n,amp}}}$ and $\overline{v^2_{\mathrm{n,amp}}}$. We can calculate the equivalent input noise current spectral density as follows:

$$\overline{i^2_{\mathrm{n,in}}} = \overline{i^2_{\mathrm{n,amp}}} + \frac{\overline{v^2_{\mathrm{n,amp}}}}{\left|\frac{R_{\mathrm{F}}}{1+\mathrm{j}\omega C_{\mathrm{T}} R_{\mathrm{F}}}\right|^2} + \overline{i^2_{\mathrm{n,R}}}, \qquad (7.50)$$

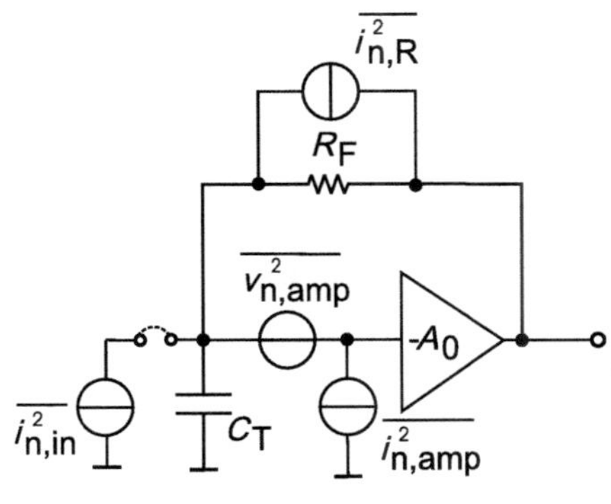

Fig. 7.13. Basic circuit of a TIA including general noise sources

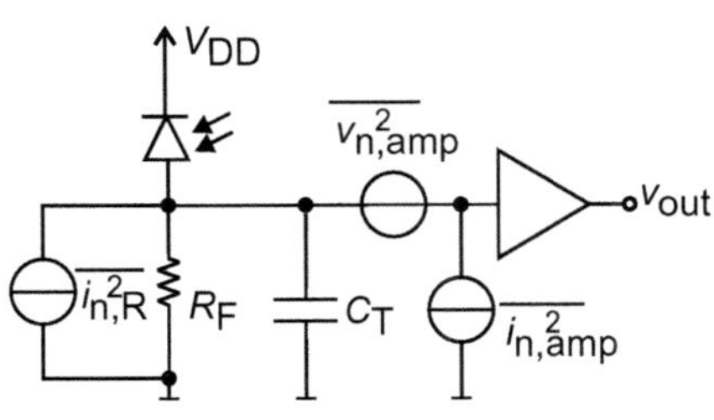

Fig. 7.14. Pin photodiode with amplifier for upper bound for noise analysis

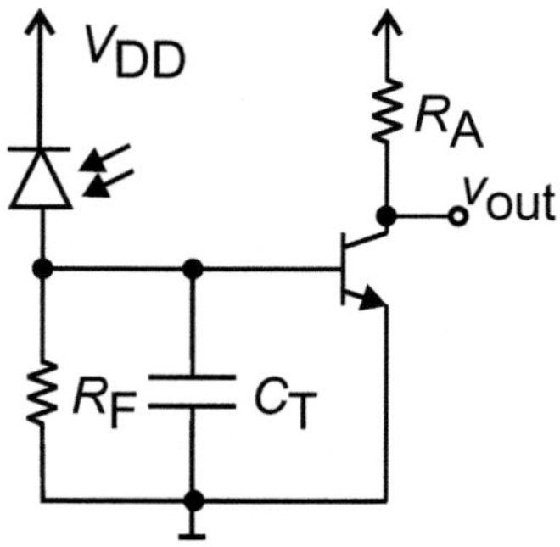

Fig. 7.15. "Small-signal" pin-BIP amplifier for noise estimation

which leads to

$$\overline{i^2_{\mathrm{n,in}}} = \overline{i^2_{\mathrm{n,amp}}} + \overline{v^2_{\mathrm{n,amp}}}\frac{\left(1 + 4\pi^2 f^2 C_{\mathrm{T}}^2 R_{\mathrm{F}}^2\right)}{R_{\mathrm{F}}^2} + \overline{i^2_{\mathrm{n,R}}}. \tag{7.51}$$

Equation (7.51) shows that the equivalent input noise current spectral density rises proportional to C_{T}^2 and f^2 at high frequencies. The total input noise current is obtained by integrating (7.51) over the bandwidth.

7.3.2 TIA with Bipolar and Heterojunction Bipolar Input Stage

For bipolar and heterojunction bipolar input stages we assume the circuit shown in Fig. 7.15.

The noise sources of the amplifier of this circuit are equal to the equivalent input noise sources of the bipolar transistor $\overline{v^2_{\mathrm{n,Ti}}}$ and $\overline{i^2_{\mathrm{n,Ti}}}$. The noise of the load resistor R_{A} is attenuated by the gain of the bipolar transistor and therefore at first approximation neglected. Equation (7.51) leads with the equivalent input noise sources of the bipolar transistors, see (7.41) and (7.36), to

$$\frac{\overline{i^2_{\mathrm{n,in}}}}{\Delta f} = \frac{4k_{\mathrm{B}}T}{R_{\mathrm{F}}} + 2q\left(\frac{I_{\mathrm{C}}}{\beta} + \frac{K_1}{(2q)^2}\left(\frac{I_{\mathrm{C}}}{\beta}\right)^a + \frac{I_{\mathrm{C}}}{\left|\frac{\beta}{1+\mathrm{j}f/f_\beta}\right|^2}\right) + \frac{4\left(1 + 2\pi f C_{\mathrm{T}} R_{\mathrm{F}}\right)^2 k_{\mathrm{B}}T\left(r_{\mathrm{b}} + \frac{1}{2g_{\mathrm{m}}}\right)}{R_{\mathrm{F}}}. \tag{7.52}$$

The calculation for the SiGe heterojunction bipolar transistor (HBT) is in fact the same, except different values for the parameters.

7.3.3 TIA with MOS Input Stage

For a MOS-input stage the pin-amplifier circuit for upper noise boundary is shown in Fig. 7.16. This circuit is called a pin-FET amplifier [101]. Again the

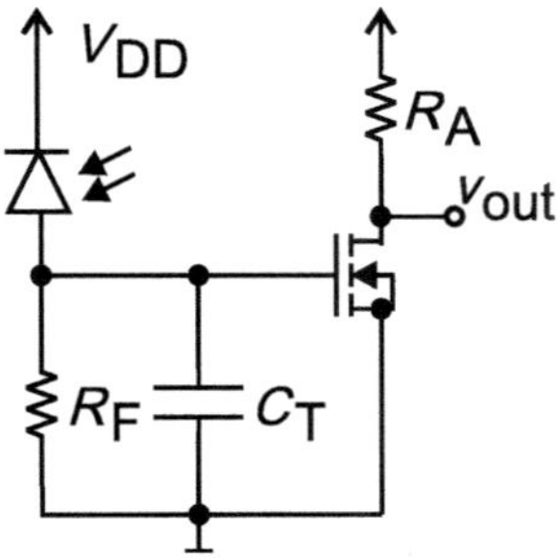

Fig. 7.16. "Small-signal" pin-FET amplifier as upper bound for FET-TIA noise analysis

noise sources of the amplifier equal the equivalent input noise sources of the FET $\overline{v^2_{\text{n,Ti}}}$ and $\overline{i^2_{\text{n,Ti}}}$. The noise of the load resistor R_A again is neglected, because it is attenuated by the transconductance of the FET, if it is referred to the input.

Substituting the noise generators shown in (7.46) and (7.49) leads to:

$$\overline{i^2_{\text{n,in}}} = \frac{4k_\text{B}T}{R_\text{F}} + \frac{4\pi^2 f^2 C_\text{T}^2}{g_\text{m}^2}\left(4k_\text{B}T\frac{2}{3}g_\text{m} + K_\text{f}\frac{I_\text{D}^a}{f}\right)$$
$$+\frac{1+4\pi^2 f^2 C_\text{T}^2 R_\text{f}^2}{R_\text{f}^2 g_\text{m}^2}\left(4k_\text{B}T\frac{2}{3}g_\text{m} + K_\text{f}\frac{I_\text{D}^a}{f}\right) + 2qI_\text{D}. \tag{7.53}$$

The average optical input power P_opt is calculated by (7.54):

$$P_\text{opt} = \frac{\hbar\omega}{\eta q}\sqrt{\int_B \overline{i^2_{\text{n,in}}}}. \tag{7.54}$$

Figure 7.17 shows the sensitivity of the pin-FET amplifier over the bandwidth and for different wavelengths.

Due to the fact that the bandwidth is essentially dependent on the feedback resistance as shown in (7.55), R_F is varied in Fig. 7.17 in dependence on the bandwidth assuming constant A_0 and C_T

$$f_{-3\,\text{dB}} = \frac{A_0 + 1}{2\pi R_\text{F} C_\text{T}}. \tag{7.55}$$

In the example depicted in Fig. 7.17 a value for R_F of 3.2 kΩ gives a bandwidth of 90 MHz with the pin-FET amplifier,which has a A_0 of 0.

7.3.4 Comparison of Bipolar and Field-Effect Transistor Circuits Based on Noise Theory

The comparison of bipolar and field-effect transistor circuit will be based on the pin-FET or pin-BJT amplifier circuit shown in Figs. 7.15 and 7.16.

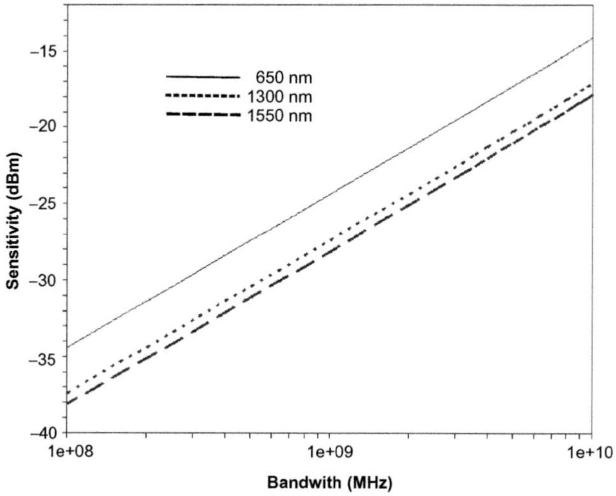

Fig. 7.17. Sensitivity of pin-FET amplifier for different wavelengths and variable R_F according to (7.55), BER= 10^{-9}, $g_m = 10\,\text{mS}$, $C_T = 0.5\,\text{pF}$, $\eta = 0.5$, $EX = \infty$

The key features of the circuits are summarized later. We obtain for both circuits the same feedback resistor R_F and the same input node capacitance ($C_T = 1\text{pF}$), and the flicker noise is neglected. R_F is assumed to be dependent on the $-3\,\text{dB}$ cut-off frequency f_g according to $R_F = 1/(2\pi f_g C_T)$ for the simple pin-amplifier configuration. The $-3\,\text{dB}$ cut-off frequency is estimated to be 2/3 of the data rate for optimum noise behavior [7]. Therefore the value of R_F can be easily calculated for each data rate with a given input-node capacitance. The temperature T is assumed to be about room temperature ($T = 300\,\text{K}$):

$$R_F = 1/(2\pi f_g C_T)$$

We obtain for the Si-bipolar transistor the cut-off frequency of β f_β, the collector current I_C, the base resistance r_b, and the transconductance $g_{m_{BIP}}$ with the values shown in Table 7.2. The SiGe HBT shows a smaller base resistance r_b, a higher β as well as a higher cut-off frequency of f_β (see Table 7.2).

The transconductance $g_{m_{FET}}$ and the drain current I_D of the FET are given in Table 7.2. Flicker noise and the noise of the gate current I_G are neglected.

Figure 7.18 shows the equivalent input noise current for Si bipolar, MOSFET, and SiGe HBT circuits. The calculation was done with typical values for a 0.6 μm Si bipolar, a 0.12 μm CMOS and a 0.35 μm SiGe HBT process. At low frequencies (≤100 MHz) the bipolar transistor generates a higher equivalent input noise current due to the base current and the base resistor r_b. It can be seen at frequencies in the GHz-range that deep-sub-micron CMOS

Table 7.2. Example values for Si-bipolar and MOSFET amplifier

Si-bipolar amplifier	deep-sub-micron MOSFET amplifier	SiGe HBT
$\beta_{\mathrm{Si}} = 80$	$C_{\mathrm{ox}} = 10\,\mathrm{fF\,\mu m^{-2}}$	$\beta_{\mathrm{SiGe}} = 160$
$f_{\beta,\mathrm{Si}} = 500\,\mathrm{MHz}$	$W = 50\,\mathrm{\mu m}$	$f_{\beta,\mathrm{SiGe}} = 5\,\mathrm{GHz}$
$r_{\mathrm{b,Si}} = 300\,\Omega$	$L = 0.12\,\mathrm{\mu m}$	$r_{\mathrm{b,SiGe}} = 50\,\Omega$
$g_{\mathrm{m,BIP,Si}} = 7.7\,\mathrm{mS}$	$g_{\mathrm{m_{FET}}} = 6.7\,\mathrm{mS}$	$g_{\mathrm{m,BIP,SiGe}} = 7.7\,\mathrm{mS}$
$I_{\mathrm{C,Si}} = 200\,\mathrm{\mu A}$	$I_{\mathrm{D}} = 200\,\mathrm{\mu A}$	$I_{\mathrm{C,SiGe}} = 200\,\mathrm{\mu A}$

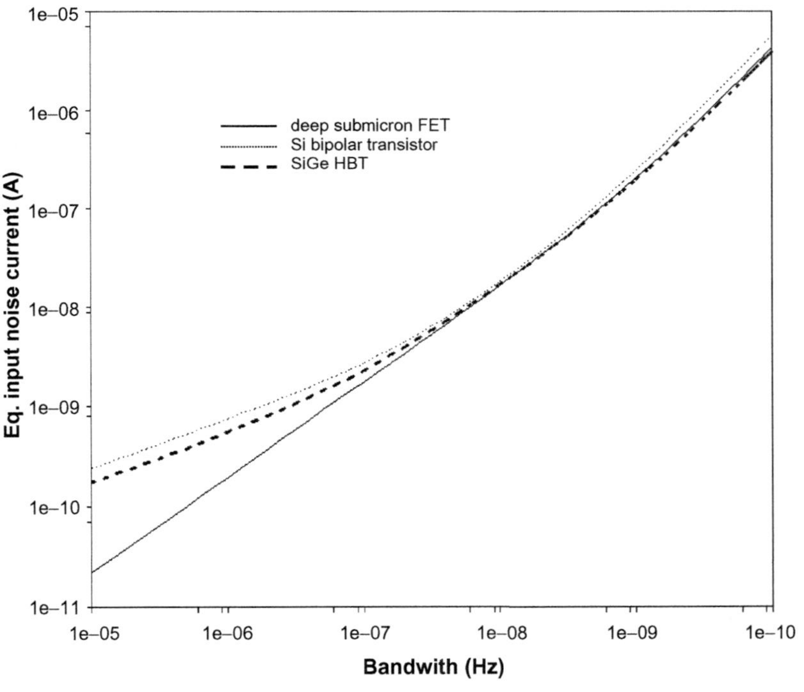

Fig. 7.18. Input noise current according to (7.52) for Si and SiGe bipolar and (7.53) for MOSFET pin-amplifier circuit over the bandwidth, with values of Table 7.2

has advantages over the 0.6 μm Si bipolar process, but the 0.35 μm SiGe HBT already shows a slight advantage over the deep-sub-micron CMOS, although the minimum structure size is larger than in the CMOS process. Figure 7.19 shows that the deep-sub-micron FET leads to a better sensitivity than the 0.6 μm Si BJT. The 0.35 μm SiGe HBT is only slightly better for data rates $>1\,\mathrm{Gb\,s^{-1}}$ than the deep-sub-micron FET. It can be expected that also Si bipolar processes with smaller minimum structure size show a better noise behavior than the 0.12 μm CMOS technology.

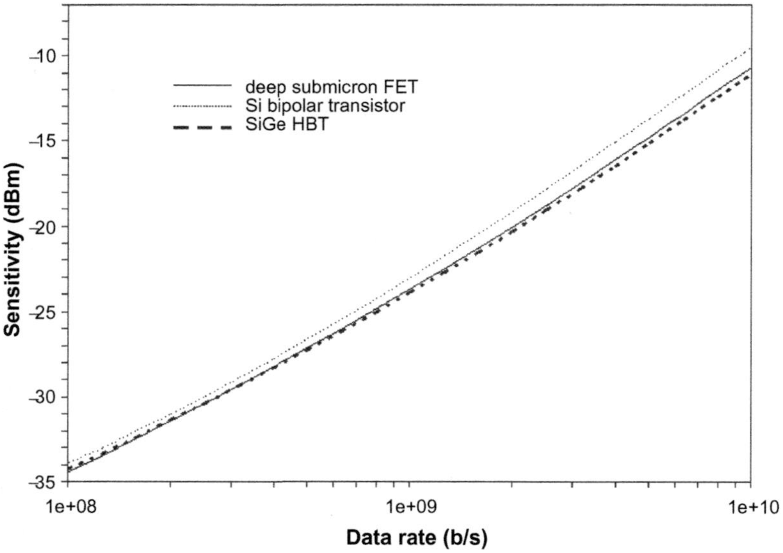

Fig. 7.19. Sensitivity comparison between pin-amplifier circuit with bipolar, MOS-FET, and SiGe-HBT amplifier

7.4 Noise Models of More Complex TIAs in Deep-Sub-micron CMOS Technology

In Sect. 7.4.1, more complex TIA topologies are discussed concerning noise. Starting with noise analysis of the folded-cascode TIA, in Sect. 7.4.2 the noise analysis of a three-stage inverter TIA is presented. In the end of the section a simplification for a realistic noise estimation of a three-stage TIA is described.

7.4.1 Noise Analysis of Folded-Cascode TIA

Figure 7.20 shows the basic circuit of the folded-cascode transimpedance amplifier.

For the calculation of the equivalent input noise current density $\overline{i^2_{\mathrm{n,in}}}$ of the TIA, the equivalent input noise sources $\overline{i^2_{\mathrm{in,amp}}}$ and $\overline{v^2_{\mathrm{in,amp}}}$ of the folded-cascode *amplifier* stage are calculated first.

Noise of the Folded-Cascode Amplifier

The equivalent input noise sources of the folded-cascode (FC) amplifier will be derived first. The current sources I1 and I2 in Fig. 7.20 are realized as MOS-transistors M3 and M4 shown in the small-signal equivalent circuit (see Fig. 7.21a). The mesh and node equations of the small-signal equivalent circuit of the folded cascode are discussed in Sect. 6.3.1. Additionally the noise current sources i_{d1} to i_{d4} are included compared to Fig. 6.12.

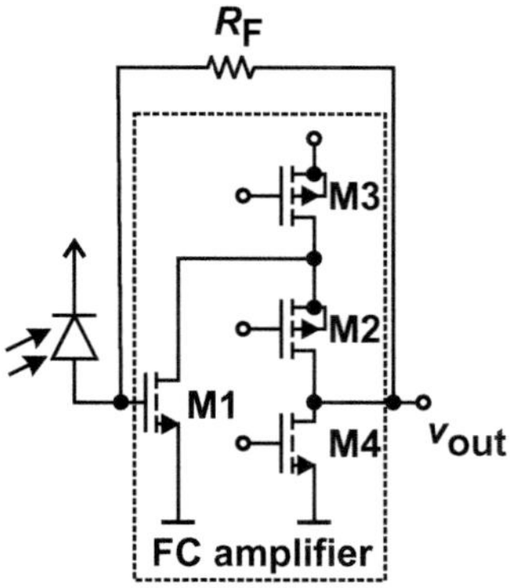

Fig. 7.20. Basic circuit of folded-cascode input stage

For the calculation of the equivalent input noise voltage density the inputs of both circuits in Fig. 7.21 are short circuited. This leads to the node equations (7.56) and (7.57) instead of (6.36) and (6.37)

$$0 = i_3 + i_{c3} + i_1 + i_{d1} + i_{d3} - i_{d2} - i_2, \tag{7.56}$$

$$0 = +i_2 + i_{d2} + i_o + i_4 + i_{d4} + i_{c4}. \tag{7.57}$$

For short-circuited output node the output current i_o is first derived for the individual noise sources of the transistors i_{d1}–i_{d4}.

This leads to an output noise current density $\overline{i^2_{out,id}}/\Delta f$ being characteristic over the bandwidth as shown in Fig. 7.22.

This output noise current density $\overline{i^2_{out,id}}/\Delta f$ is calculated with the example data of Table 6.1.

For the calculation of the equivalent input noise voltage density $\overline{v^2_{in,amp}/\Delta f}$ all devices are assumed as not noisy and the output current $i_{out,v}$ due to the equivalent input noise voltage $v_{in,amp}$ is calculated again.

Equation (7.58) is inserted into the standard mesh and node equations of the folded cascode derived in Sect. 6.3.1.

$$v_{in,amp} = v_{in}. \tag{7.58}$$

The resulting output current $i_{out,v}$ due to the equivalent input noise voltage is equated to the output current $i_{out,id}$ due to the individual noise sources of the transistors (see (7.59)).

$$i_{out,v}(v_{in,amp}) = i_{out,id}(i_{d1}, i_{d2}, i_{d3}, i_{d4}). \tag{7.59}$$

$\overline{v^2_{in,amp}/\Delta f}$ is calculated in dependence on i_{d1} to i_{d4} with short circuited output and short circuited input, equivalent to the calculation of the equivalent input noise voltage density of a single transistor. The frequency-dependent input noise voltage density is displayed in Fig. 7.24.

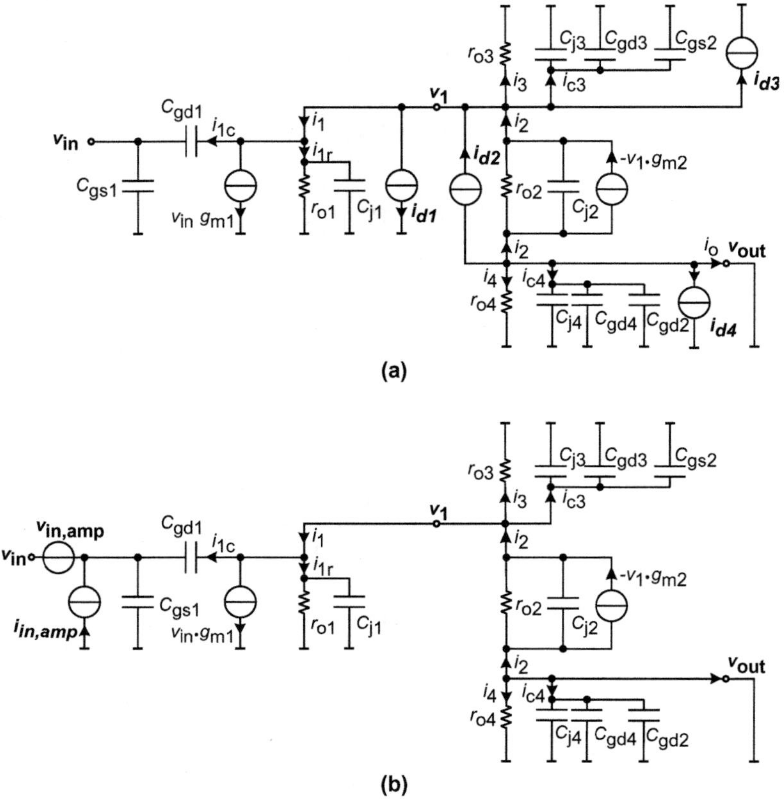

Fig. 7.21. (**a**) Small-signal equivalent circuit including noise sources; (**b**) small-signal equivalent circuit with equivalent input noise sources

For the calculation of the equivalent input noise current density $\overline{i_{\mathrm{in,amp}}^2/\Delta f}$ the input node is open circuited in both circuits of Fig. 7.21. Again the output current $i_{\mathrm{out,id}}$ is calculated; this time the input voltage v_{in} is determined by the current i_{1c} (see (7.60)):

$$v_{\mathrm{in}} = \frac{i_{\mathrm{1c}}}{\mathrm{j}2\pi f C_{\mathrm{gs1}}}. \tag{7.60}$$

Afterwards the output current due to the input noise current source $i_{\mathrm{out,iin}}$ is calculated. The input node is open circuited and therefore the v_{in} is given by

$$v_{\mathrm{in}} = \frac{i_{\mathrm{in}} + i_{\mathrm{1c}}}{j2\pi f C_{\mathrm{gs1}}}. \tag{7.61}$$

The two calculated output currents $i_{\mathrm{out,id}}$ and $i_{\mathrm{out,iin}}$ are equated and the equivalent input noise current density of the folded-cascode amplifier is derived

$$i_{\mathrm{out,i}}(i_{\mathrm{in,amp}}) = i_{\mathrm{out,id}}(i_{\mathrm{d1}}, i_{\mathrm{d2}}, i_{\mathrm{d3}}, i_{\mathrm{d4}}). \tag{7.62}$$

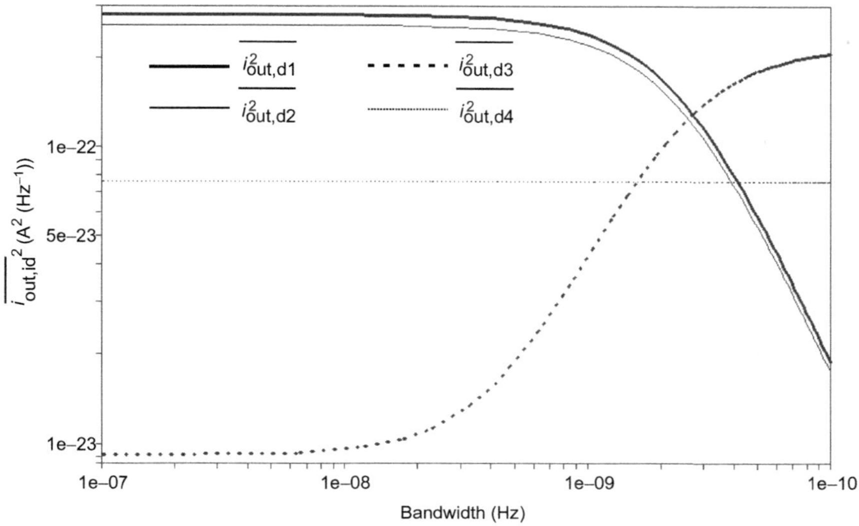

Fig. 7.22. Output noise current density over the bandwidth, depending on i_{d1}–i_{d4}

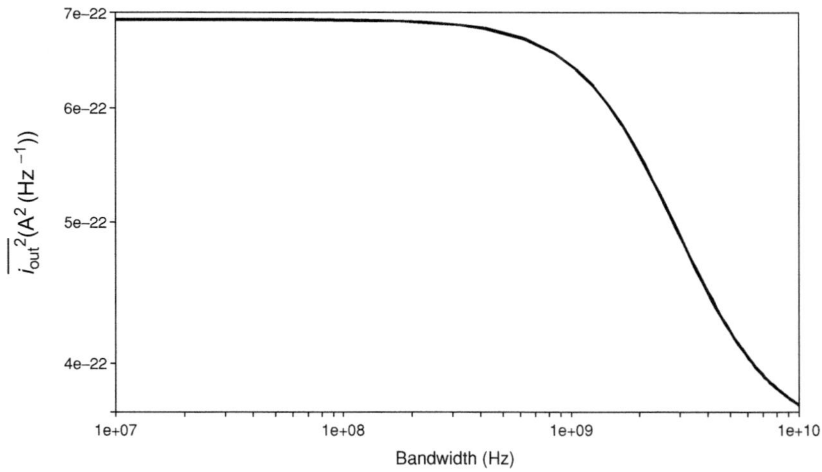

Fig. 7.23. Summarized output noise current density $\overline{i^2_{\text{out}}}$ over the bandwidth

The equivalent input noise current density of the folded-cascode amplifier $\overline{i^2_{\text{in,amp}}}$ is displayed in Fig. 7.25.

Folded-cascode TIA

Now we add R_{F} to the circuit shown in Fig. 7.21. The input noise current density of the folded-cascode TIA $\overline{i^2_{\text{n,in}}}$ is calculated the same way as discussed in Sect. 7.3. As input noise sources of the amplifier the noise sources of the folded-cascode amplifier are inserted.

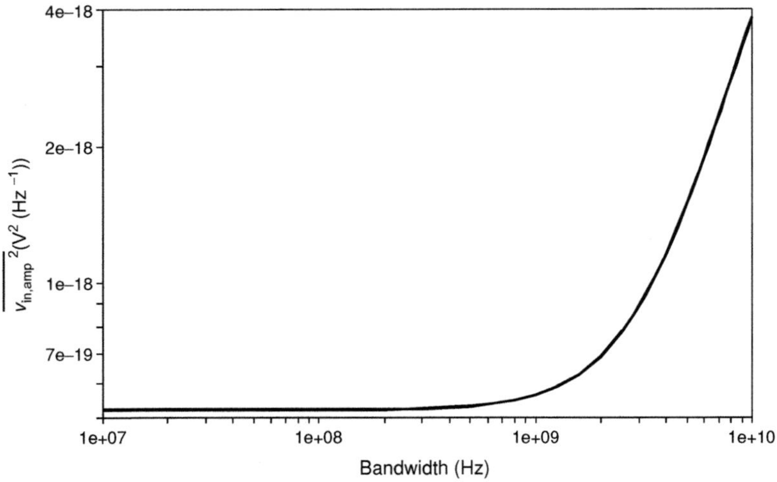

Fig. 7.24. Equivalent input noise voltage density $\overline{v_{\text{in,amp}}^2}$ of folded-cascode amplifier

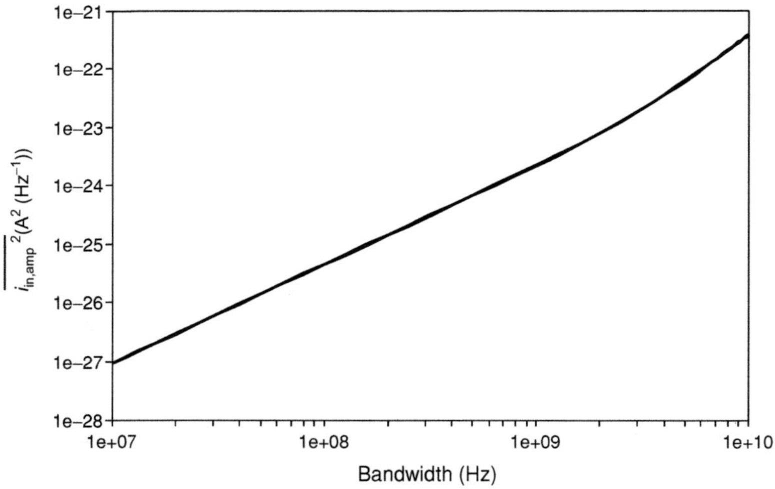

Fig. 7.25. Equivalent input noise current density $\overline{i_{\text{in,amp}}^2}$ of folded-cascode amplifier

The following calculations are done with $R_{\text{F}} = 1,800\,\Omega$, $C_{\text{F}} = 80\,\text{fF}$, and $C_{\text{pd}} = 1.0\,\text{pF}$.

The equivalent input noise current density of the folded-cascode TIA is displayed in Fig. 7.26.

Integration of the equivalent input noise current density of the folded-cascode TIA over the bandwidth gives the equivalent input noise current $\overline{I_{\text{n,in}}^2}$. For the presented example $\overline{I_{\text{n,in}}^2} = 2.539 \times 10^{-14}\,\text{A}^2$ where the bandwidth of 830 MHz for a data rate of $1\,\text{Gb}\,\text{s}^{-1}$ is taken. This corresponds to an

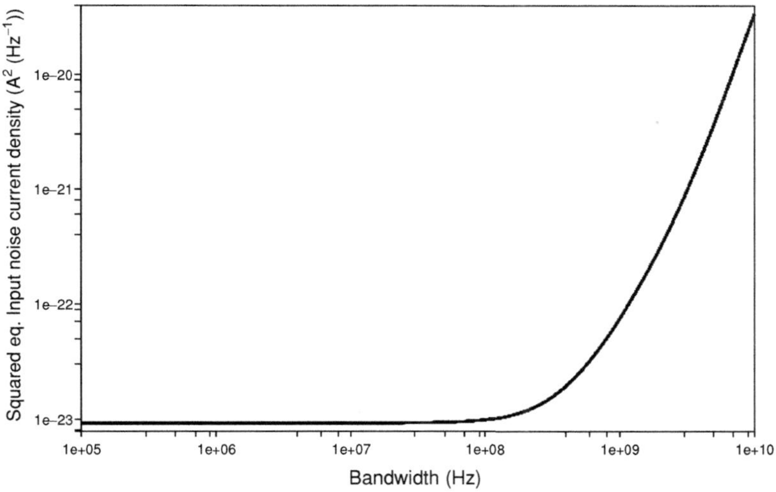

Fig. 7.26. Equivalent input noise current density $\overline{i^2_{\mathrm{n,in}}}$ of folded-cascode TIA

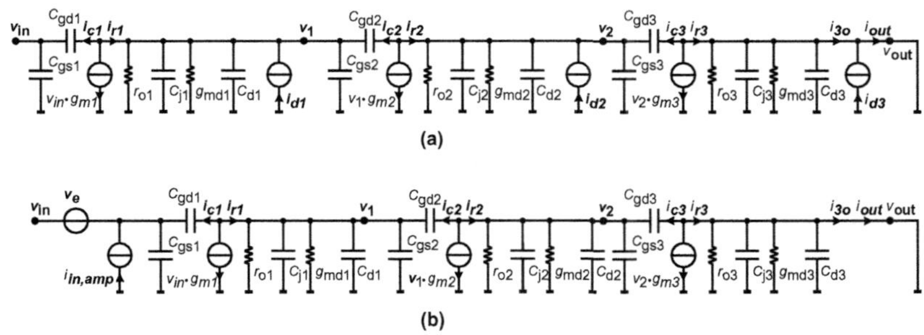

Fig. 7.27. Equivalent small-signal circuit including noise sources of each stage (**a**) and including the equivalent input noise sources of the three-stage amplifier (**b**)

rms 160 nA equivalent input noise current and leads to a sensitivity of about -28 dBm with $R = 0.85\,\mathrm{A\,W^{-1}}$, EX = 10 and BER = 10^{-10}. This meets the results of the circuit simulation quite well.

7.4.2 TIA with CMOS-Inverter Input Circuit

Now, the noise calculation of the three-inverter TIA shown in Fig. 6.16 will be discussed. For the calculation of the equivalent input noise current density, again the input noise sources of the amplifier will be calculated.

Figure 7.27 shows the equivalent small-signal circuit including the noise sources of each stage. Again it should be mentioned that the elements of the circuit represent the elements of the whole stage, e.g., $g^{\mathrm{I}}_{\mathrm{m1}} = g_{\mathrm{mn1}} + g_{\mathrm{mp1}}$ for the transconductances of the transistors N1 and P1. Also the inserted noise sources represent the noise of the whole stage:

$$\frac{\overline{i_{\mathrm{I,d1}}^2}}{\Delta f} = 4k_{\mathrm{B}}T(g_{\mathrm{mn1}} + g_{\mathrm{mp1}} + g_{\mathrm{md1}}) \tag{7.63}$$

with the transconductances of the transistors N1, P1 and the transistor of the diode load D1. T represents the temperature and k_{B} Boltzmann's constant.

The noise of the second stage

$$\frac{\overline{i_{\mathrm{I,d2}}^2}}{\Delta f} = 4k_{\mathrm{B}}T(g_{\mathrm{mn2}} + g_{\mathrm{mp2}} + g_{\mathrm{md2}}) \tag{7.64}$$

and of the third stage

$$\frac{\overline{i_{\mathrm{I,d3}}^2}}{\Delta f} = 4k_{\mathrm{B}}T(g_{\mathrm{mn3}} + g_{\mathrm{mp3}} + g_{\mathrm{md3}}) \tag{7.65}$$

are calculated equivalently.

For the calculation of the equivalent input noise voltage density $\overline{v_{\mathrm{in,amp}}^2}/\Delta f$ of the three-stage amplifier both circuits of Fig. 7.27 are short circuited and the output noise current is calculated for each current and equated. Again, $\overline{v_{\mathrm{in,amp}}^2}/\Delta f$ is calculated in dependence on the noise sources of the three stages.

$$i_{\mathrm{I,out,v}}(v_{\mathrm{I,in,amp}}) = i_{\mathrm{I,out,id}}(i_{\mathrm{I,d1}}, i_{\mathrm{I,d2}}, i_{\mathrm{I,d3}}). \tag{7.66}$$

The equivalent input noise current density $\overline{i_{\mathrm{I,in,amp}}^2}/\Delta f$ of the three-stage amplifier is calculated by open circuiting the two circuits of Fig. 7.27. Again the output currents for both circuits are calculated and equated. $\overline{i_{\mathrm{I,in,amp}}^2}/\Delta f$ depending on the noise sources of the three stages is calculated.

$$i_{\mathrm{I,out,iin}}(i_{\mathrm{I,in,amp}}) = i_{\mathrm{out,id}}(i_{\mathrm{I,d1}}, i_{\mathrm{I,d2}}, i_{\mathrm{I,d3}}) \,. \tag{7.67}$$

These two values are inserted into the noise calculation of the TIA, described in Sect. 7.3. The equivalent input noise current density of the TIA is shown in Fig. 7.28.

For a CMOS-inverter input circuit as shown in Fig. 6.16 the input referred noise current density $\overline{i_{\mathrm{in}}^2}/\Delta f$ mainly depends on the noise of the transistors of the first stage N_1 and P_1 as well as the noise of the feedback resistor R_{F}. Figure 7.29 shows clearly that the feedback resistor is the main noise source for low frequencies, with $1/f$ noise neglected, while the equivalent input noise current density at high frequencies is dominated by the noise of the first amplifier stage. All calculations are done with the data summarized in Table 6.3.

It is therefore tolerable to reduce the circuit for rough noise calculations to the circuit shown in Fig. 7.30. This circuit equals for noise analysis the pin-FET receiver shown in Fig. 7.16.

The equivalent input referred noise current density is given by (7.68).

$$\frac{\overline{i_{\mathrm{in}}^2}}{\Delta f} = \left[\left(\overline{v_{\mathrm{n,Ti,n}}^2} + \overline{v_{\mathrm{n,Ti,p}}^2}\right)\frac{1 + (\omega C_{\mathrm{T}} R_{\mathrm{F}})^2}{R_{\mathrm{F}}^2} + \overline{i_{\mathrm{n,Ti,n}}^2} + \overline{i_{\mathrm{n,Ti,p}}^2} + \overline{i_{\mathrm{R}}^2}\right] \,. \tag{7.68}$$

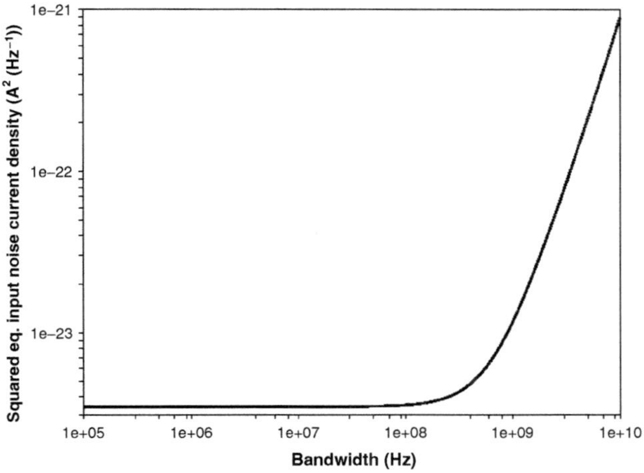

Fig. 7.28. Equivalent input noise current density of three-inverter TIA

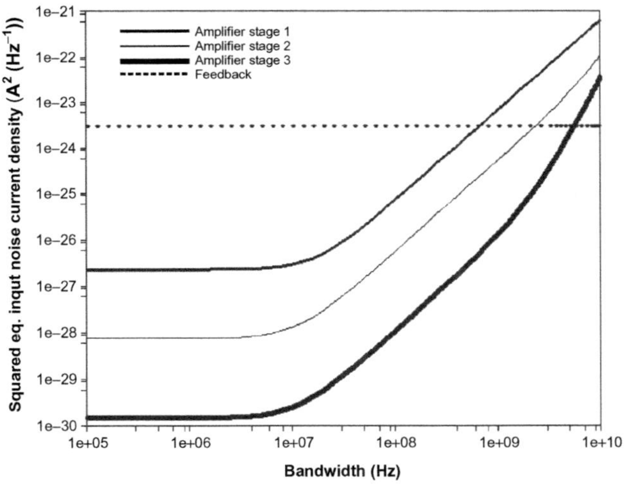

Fig. 7.29. Equivalent input noise current density due to single amplifier stages and feedback resistance

The noise of the feedback resistor $\overline{i_{\mathrm{R}}^2}$ is calculated as $\overline{i_{\mathrm{R}}^2} = 4k_{\mathrm{B}}T\frac{1}{R_{\mathrm{F}}}\Delta f$. The equivalent input noise sources of the transistors are shown in (7.70)–(7.72). These noise sources are referred to the input of the transistors, not to the TIA input!

The equivalent input noise current density of the transistor P_1 is

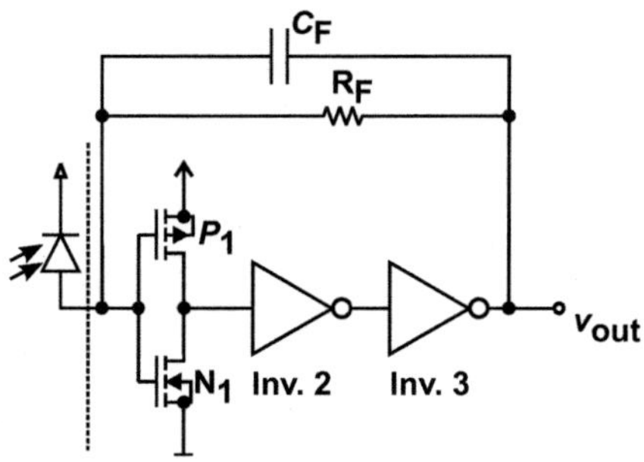

Fig. 7.30. Basic circuit of inverter input stage

$$\frac{\overline{i_{\mathrm{n,Ti,p}}^2}}{\Delta f} = 2qI_G + \omega^2 C_{\mathrm{T}}^2 \cdot \left(4k_{\mathrm{B}}T\Gamma_{\mathrm{F}} g_{\mathrm{mp}} \frac{1}{(g_{\mathrm{mp}}+g_{\mathrm{mn}})^2} + K_{1_{\mathrm{PMOS}}} \frac{I_{\mathrm{D}}^a}{(g_{\mathrm{mp}}+g_{\mathrm{mn}})^2 f} \right) . \tag{7.69}$$

The noise generated by the drain–source current generator $\overline{i_d^2}$ of one transistor (see (7.42)), in this case P_1, is transferred to the input of the inverter by the transconductances of both transistors, N_1 with g_{mn} and P_1 with g_{mp}, and therefore the equivalent input noise current density $\overline{i_{\mathrm{n,Ti,p}}^2}$ is divided by the squared sum of g_{mn} and g_{mp}.

The equivalent input noise current density of the transistor N_1 is calculated by

$$\frac{\overline{i_{\mathrm{n,Ti,n}}^2}}{\Delta f} = 2qI_{\mathrm{G}} + \omega^2 C_{\mathrm{T}}^2 \cdot \left(4k_{\mathrm{B}}T\Gamma_{\mathrm{F}} g_{\mathrm{mn}} \frac{1}{(g_{\mathrm{mp}}+g_{\mathrm{mn}})^2} + K_{1_{\mathrm{NMOS}}} \frac{I_{\mathrm{D}}^a}{(g_{\mathrm{mp}}+g_{\mathrm{mn}})^2 f} \right) . \tag{7.70}$$

Neglecting I_{G}, the equivalent noise input voltage densities of P_1 and N_1, respectively, are given by (7.71) and (7.72).

Again the calculations are done like those shown in (7.44)–(7.46). Again the noise current due to $\overline{i_d^2}$ is transferred by the sum of g_{mn} and g_{mp}

$$\frac{\overline{v_{\mathrm{n,Ti,p}}^2}}{\Delta f} = 4k_{\mathrm{B}}T\Gamma_{\mathrm{F}} \left(g_{\mathrm{mp}} \frac{1}{(g_{\mathrm{mp}}+g_{\mathrm{mn}})^2} + K_{1_{\mathrm{PMOS}}} \frac{I_{\mathrm{D}}^a}{(g_{\mathrm{mp}}+g_{\mathrm{mn}})^2 f} \right) , \tag{7.71}$$

$$\frac{\overline{v_{\mathrm{n,Ti,n}}^2}}{\Delta f} = 4k_{\mathrm{B}}T\Gamma_{\mathrm{F}} \left(g_{\mathrm{mn}} \frac{1}{(g_{\mathrm{mp}}+g_{\mathrm{mn}})^2} + K_{1_{\mathrm{NMOS}}} \frac{I_{\mathrm{D}}^a}{(g_{\mathrm{mp}}+g_{\mathrm{mn}})^2 f} \right) . \tag{7.72}$$

Due to the high bandwidth the flicker noise component is neglected in further calculations.

Substituting (7.70)–(7.72) in (7.68) and integrating over the bandwidth leads to

$$\overline{I_{\text{in}}^2} = 4k_{\text{B}}TB\left[\frac{1}{R} + 2\frac{\Gamma_{\text{F}}}{g_{\text{mp}} + g_{\text{mn}}}\frac{(2\pi C_{\text{T}}B)^2}{3} + \frac{\Gamma_{\text{F}}}{R^2\left(g_{\text{mp}} + g_{\text{mn}}\right)}\right]. \qquad (7.73)$$

$\overline{I_{\text{in}}^2}$ has a magnitude of $0.8 \times 10^{-14}\,\text{A}^2$ for a bandwidth of 830 MHz which meets the circuit simulation quite well. The rms equivalent input noise current I_{in} is the root of $\overline{I_{\text{in}}^2}$ shown in (7.73). The rms equivalent input noise current is therefore 0.9×10^{-7} A or 90 nA. This is almost half of the rms equivalent input noise current of the *folded-cascode TIA*.

8

State of the Art

This chapter gives a short overview of the state of the art concerning highly sensitive continuous-mode (CM) optical receivers (ORs) and burst-mode receivers. The main characteristic data are summarized and will be compared with our own work described in Chap. 9. The used wavelengths of the mentioned works are often not given in the original articles and therefore not known. Whenever the wavelength is given it is also mentioned here. All values of sensitivity not given in dBm are calculated to dBm with the parameter of the photodiode used in our own designs. The responsivity of the photodiode is $0.85\,\mathrm{A\,W^{-1}}$ at $\lambda = 1.3\,\mu\mathrm{m}$, the bit-error rate (BER) is 10^{-10} and the extinction ratio EX equals 10.

8.1 Silicon Bipolar and BiCMOS Optical Receivers

In [105], a CM OR in $0.6\,\mu\mathrm{m}$ BiCMOS was presented for a data rate of $622\,\mathrm{Mb\,s^{-1}}$. The depicted sensitivity was $-29.4\,\mathrm{dBm}$ with a BER $= 10^{-10}$. Maximum input power was 0 dBm and the power consumption was given with 220 mW with a supply voltage of 3.3 V.

An ac-coupled system, depicted in Fig. 8.1, was presented in [4]. This receiver uses standard modules and combines them to a burst-mode system. To avoid problems with long "1" or "0" bit sequences due to the ac-coupling a 8 B/10 B coding is assumed to be transmitted. This means that every 8-bit sequence is coded to 10 bits. The photodiode-preamplifier module is an InGaAs pin-TIA module (Philips, TZA3043) which shows a sensitivity of $-27.7\,\mathrm{dBm}$ at $1.25\,\mathrm{Gb\,s^{-1}}$ with a PRBS of $2^9 - 1$ and a BER $= 10^{-12}$. The preamplifier itself is presumably produced in Si bipolar technology [106]. The measured switching time for a dynamic range of 21 dB is 75 ns.

A burst-mode receiver was presented in [107]. This design combines an external InGaAs avalanche photodiode with an Si bipolar TIA. At a data rate of $1.5\,\mathrm{Gb\,s^{-1}}$ the BMR achieves a sensitivity of $-34\,\mathrm{dBm}$ with a BER of 10^{-10}. The multiplication factor of the APD is 10 for the detected 1,300 nm light. If

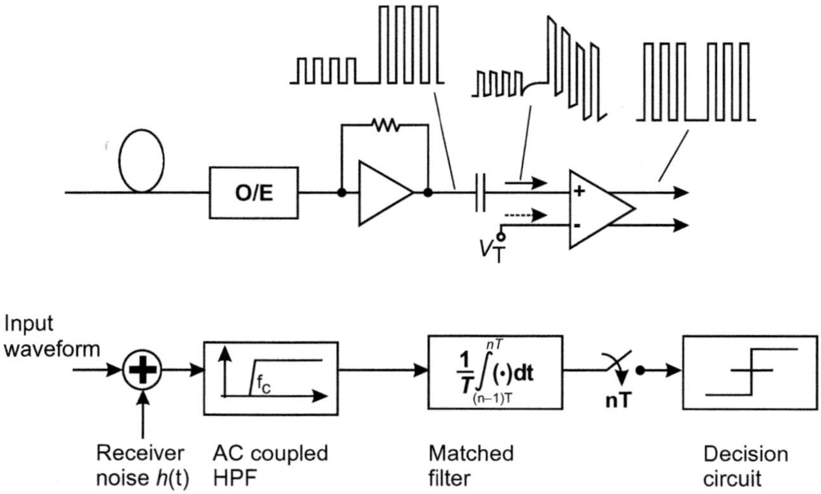

Fig. 8.1. AC-coupled OR – 8B/10B line code [4]

estimating the sensitivity of this receiver for a pin photodiode instead of the APD we obtain a value of −24 dBm. The external input node capacitance is 0.5 pF as sum of the APD and packaging. Switching between maximum and minimum optical input power is done in 50 ns for a dynamic range of 15 dB. This design is also useable in a 622 $\mathrm{Mb\,s^{-1}}$ optical bus application. Figure 8.2 shows a block diagram of the receiver for 622 $\mathrm{Mb\,s^{-1}}$ bus application. The power dissipation was 110 mW at a supply voltage of ±5 V.

In [108] a 1.7 GHz optoelectronic receiver in a 0.8 μm BiCMOS technology is presented. Noise analysis leads to an equivalent input noise current density of 9.8 $\mathrm{pA\,Hz^{-1/2}}$ which resulted in a sensitivity of −23.3 dBm at 3 $\mathrm{Gb\,s^{-1}}$ for a photodiode with an quantum efficiency of 75% at 850 nm light. The capacitance of the photodiode was assumed to be 0.5 pF [108]. In Fig. 8.3 the schematic of the TIA is presented.

A 4 × 2 $\mathrm{Gb\,s^{-1}}$ synchronous optical data link is described in [109]. The dc-coupled channels are produced in a monolithic Si bipolar IC and use an external InGaAs pin photodiode array for the detection of the 1.3 μm light. Both components are in a hybrid package. The receiver dissipates 865 mW at a supply voltage of −5.5 V. The achieved sensitivity was −12.5 dBm and −11 dBm with a BER of 10^{-9} at 1.25 and 2 $\mathrm{Gb\,s^{-1}}$, respectively. Crosstalk between the channels was given with less than 20 dB over the useful range of frequencies.

Another work [110] describes an OR module for 2.5 $\mathrm{Gb\,s^{-1}}$. It was produced in a 0.5 μm Si bipolar process technology. The optical input power range of 17.4 dB was reached without external adjustment. In Fig. 8.4 the schematic of the preamplifier is depicted. The diameter of the InGaAs/InP photodiode is large with 60 μm. The preamplifier is connected via bumps to

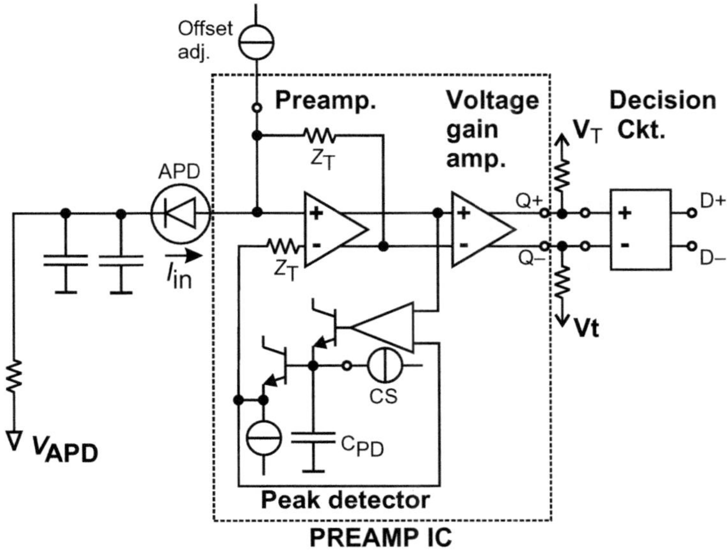

Fig. 8.2. Block diagram of the receiver for 622 Mb s^{-1} bus application [107]

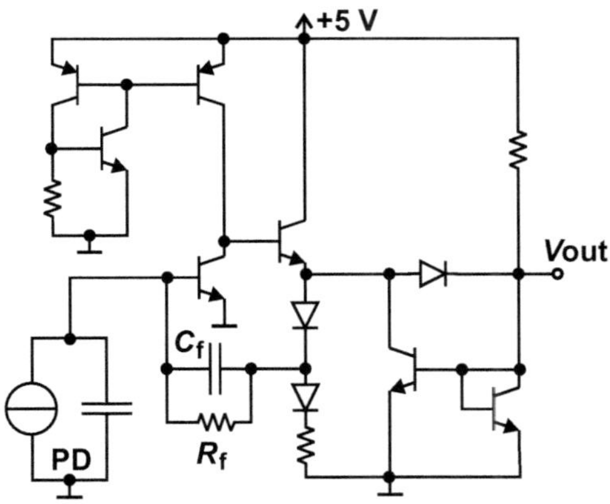

Fig. 8.3. Schematic of 1.7 GHz TIA in 0.8 μm BiCMOS technology [108]

the photodiode. The maximum sensitivity for a BER of 10^{-9} was −19.4 dBm at a bit rate of 2.5 Gb s^{-1}.

A high-gain pin-preamplifier module where the preamplifier is mounted together with the photodiode in one package was produced in an Si bipolar process [111]. The achieved sensitivities were −24 and −20 dBm at 1.9 and 3.5 Gb s^{-1}, respectively. The power dissipation was 210 mW with a supply voltage of 5 V.

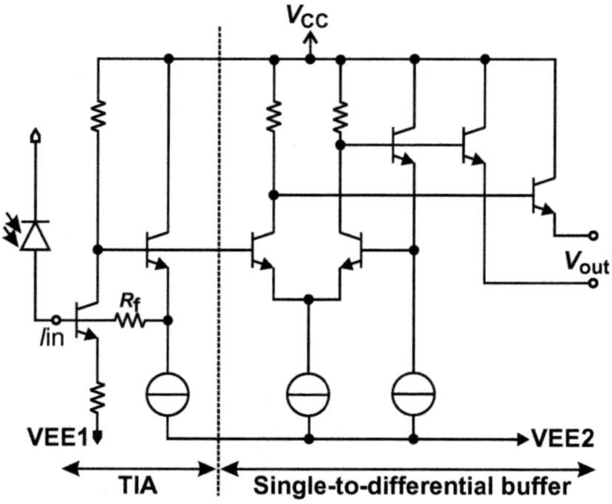

Fig. 8.4. Preamplifier circuit diagram of a low-power 2.5 Gb s^{-1} optical receiver module [110]

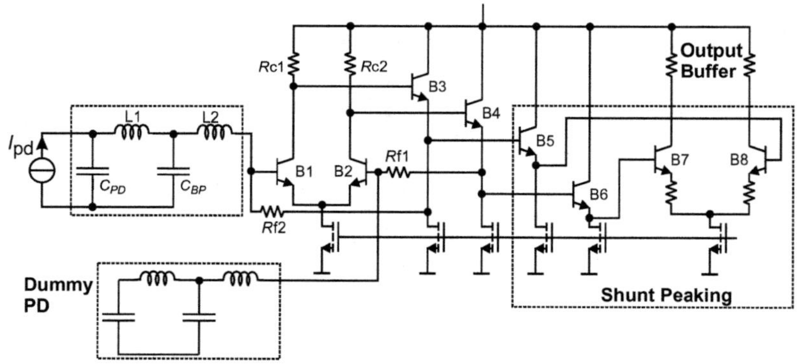

Fig. 8.5. Schematic of TIA with shunt peaking for 10 Gb s^{-1} data rate [114]

An OR with 5 Gb s^{-1} data rate in silicon bipolar was presented in [112]. The measured sensitivity was −11 dBm (BER = 10^{-12}) at 5 Gb s^{-1} with a dynamic range of 11 dB. The power consumption was given with 700 mW.

In [113, 114] a differential TIA for a 10 Gb s^{-1} synchronous optical network (SONET) receiver is presented. An InP pin photodiode was used with a 0.25 μm Si BiCMOS technology. The photodiode and a dummy photodiode were wire bonded to the preamplifier. To improve the bandwidth from 5.6 to 8.2 Gb s^{-1} the shunt peaking technique was used [114]. Figure 8.5 shows the schematic of the TIA with emitter followers B5 and B6 used as shunt peaking elements instead of the commonly used inductors and parallel capacitors. A sensitivity of −17 dBm was reached at 10 Gb s^{-1} with a BER of 10^{-12}. The power dissipation was 140 mW including 80 mW dissipation of the output buffer. The power supply voltage was 5 V.

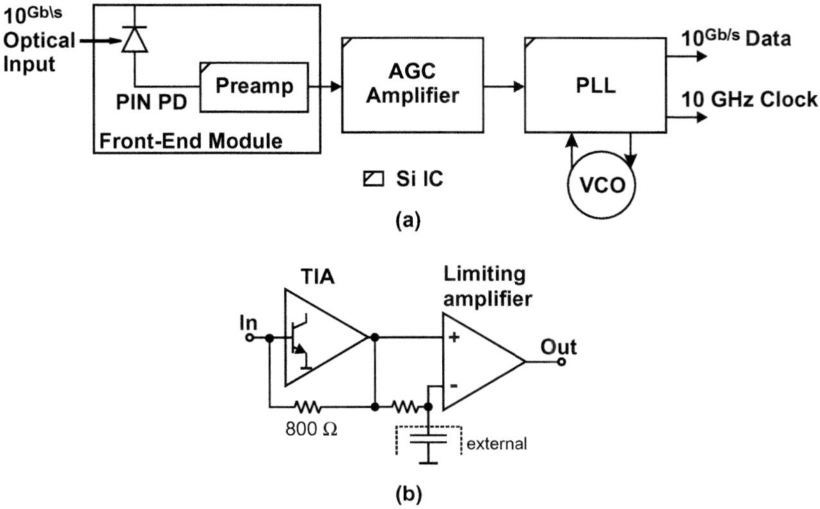

Fig. 8.6. 10 Gb s^{-1} Si receiver: (**a**) block diagram of OR; (**b**) block diagram of preamplifier [115]

Another 10 Gb s^{-1} Si receiver was presented in [115]. A 0.3 μm Si bipolar IC was combined with a pin photodiode to the receiver front-end (see Fig. 8.6a). The transimpedance is given with 800 Ω and also an external capacitance is connected to the limiting amplifier as shown in Fig. 8.6b. Without any details about the photodiode the sensitivity was given with −18.1 dBm at 10 Gb s^{-1} and a PRBS of $2^{23} - 1$. The dynamic range was about 20 dB.

In a 0.5 μm Si bipolar technology a preamplifier for optical fiber receivers was produced [116]. To simplify electrical testing a simulation circuit of the photodiode was integrated on-chip, but nevertheless connected by a bond wire to the TIA input to have the correct behavior. The on-chip inductive load L2 (see Fig. 8.7) improves the bandwidth of the TIA as well as the input bond wire inductance L1 which also improves the signal-to-noise ratio (SNR). The average input noise current density was given with 8 pA Hz$^{-1/2}$ with a photodiode capacitance of 150 fF. This value equals −21.5 dBm for the photodiode values mentioned at the beginning of the chapter at a data rate of 15 Gb s^{-1} assuming the bandwidth as 2/3 of the data rate. The power dissipation of the TIA itself was 9 mW at a supply voltage of 2.3 V.

In a 0.25 μm Si bipolar technology a 10 Gb s^{-1} OR module was designed [117]. An InGaAs pin photodiode was used to detect the 1,550 nm light. No further details of the photodiode were mentioned except the power supply voltage of 5 V for both the photodiode and the preamplifier. The achieved sensitivities were −16.4, −15.5, and −14.5 dBm for 10, 11.25 Gb and 12, respectively. The power consumption was 410 mW.

In [118] a 13 Gb s^{-1} preamplifier for optical front-ends is presented. The design was done in a 0.8 μm Si preproduction technology. In this design the

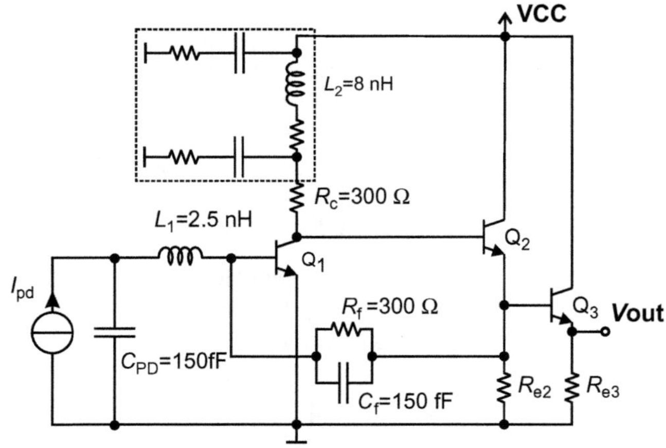

Fig. 8.7. 10 GHz Si bipolar TIA with bandwidth enhancing inductors L1 and L2 [116]

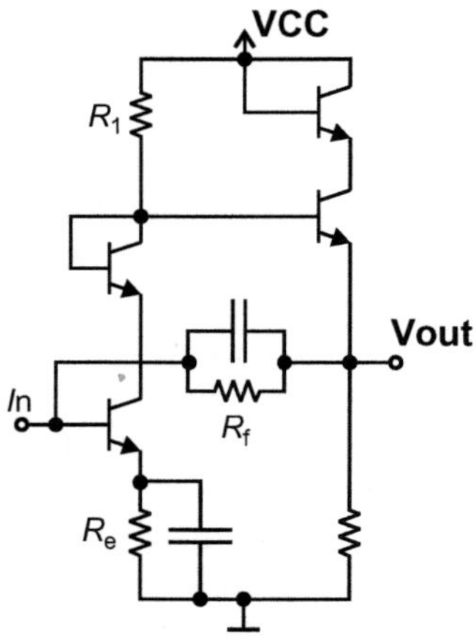

Fig. 8.8. TIA circuit of 10 Gb s^{-1} Si bipolar OR module [117]

bond inductance between external photodiode and preamplifier is used to improve the noise performance of the TIA. For experimental results a photodiode model in thin-film technology was used with a capacitance of 100 fF. Furthermore a responsivity of 1 A W^{-1} was assumed for the photodiode and therefore the estimated sensitivity was given with -22.5 dBm at 10 Gb s^{-1} for a BER of 10^{-10}.

Table 8.1 summarizes the sensitivities, power consumptions, and input power ranges of these results.

Table 8.1. Summary of sensitivities of state of the art of Si bipolar designs

reference	process	sensitivity	power cons. (mW)	in. pow. range (dB)
[105]	0.6 µm BiCMOS	−29.4 dBm @ 622 Mb s^{-1}		
[4]	Si bipolar	−27.7 dBm @ 1.25 Gb s^{-1}		21
[107]	Si bipolar	−34 dBm @ 1.5 Gb s^{-1}[a]	110	15
[108]	0.8 µm Si BiCMOS	−23.3 dBm @ 1.7 Gb s^{-1}		
[109]	Si bipolar	−12.5 dBm @ 1.25 Gb s^{-1}	865	
[109]	Si bipolar	−11 dBm @ 2 Gb s^{-1}	865	
[110]	0.5 µm Si bipolar	−19.4 dBm @ 2.5 Gb s^{-1}		17.4
[111]	Si bipolar	−24 dBm @ 1.9 Gb s^{-1}	210	
[111]	Si bipolar	−20 dBm @ 3.5 Gb s^{-1}	210	
[112]	Si bipolar	−11 dBm @ 5 Gb s^{-1}	700	
[113, 114]	0.25 µm BiCMOS	−17 dBm @ 10 Gb s^{-1}	140	
[115]	0.3 µm Si bipolar	−18.1 dBm @ 10 Gb s^{-1}		20
[116]	0.5 µm Si bipolar	−21.5 dBm @ 15 Gb s^{-1}[b]	9[c]	
[117]	0.25 µm Si bipolar	−16.4 dBm @ 10 Gb s^{-1}	41	
[117]	0.25 µm Si bipolar	−15.5 dBm @ 11.25 Gb s^{-1}	41	
[117]	0.25 µm Si bipolar	−14.5 dBm @ 12 Gb s^{-1}	41	
[118]	0.8 µm Si bipolar	−22.5 Gb s^{-1} @ 13 Gb s^{-1}	280	

[a] APD, $m = 10$

[b] Calculated value out of given input noise value and data of the photodiode described in the beginning of the chapter

[c] Preamplifier only

8.2 SiGe Heterojunction Bipolar and SiGe BiCMOS Optical Receivers

The superior speed of SiGe heterojonction bipolar transistors (HBTs) compared to Si bipolar transistors has already been mentioned and the structure of a monolithic SiGe–Si pin-HBT receiver was described in Chap. 3.4. Here, the bipolar transimpedance amplifier circuit of this receiver (see Fig. 8.9) is discussed.

The receiver consists of a pin photodiode, a common emitter gain stage, two emitter follower buffers, and a resistive feedback loop. NiCr thin-film resistors were used in the monolithic SiGe HBT receiver [77]. The transistors Q1, Q4, and Q5 are used as level-shifting diodes. Q1 and Q5 reduce U_{CE} of Q2 and Q6, respectively, because the breakdown voltages of high-speed transistors are quite low.

The two voltage sources VDD and VCC were necessary to optimize the pin transient behavior and the operating point of the amplifier. The value of the feedback resistor R_{F} determines the bandwidth, gain, and noise characteristics of the photoreceiver. The value of R_{F} is usually chosen based on a tradeoff

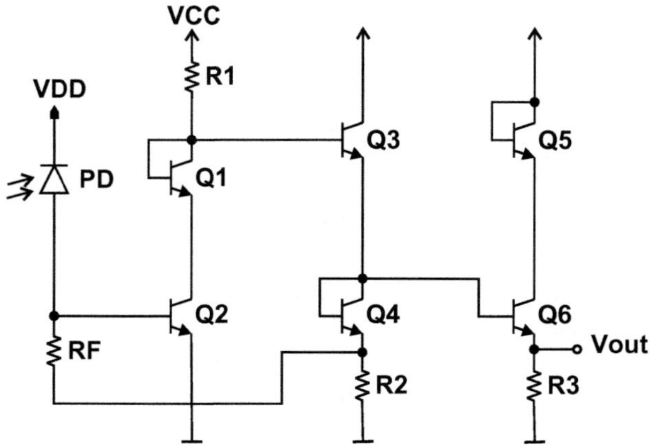

Fig. 8.9. Circuit diagram of a bipolar photoreceiver [77]

between these three parameters. In [77], a value of 640 Ω was chosen for R_F resulting in a transimpedance gain of 52.2 dB Ω. The bandwidth of 1.6 GHz was obtained for the transimpedance amplifier with an f_T of 25 GHz for the HBTs with an emitter area of 5 × 5 µm. The optical bandwidth of 460 MHz of the pin-HBT receiver was measured for VDD = 9 V and VCC = 6 V. The bandwidth of the receiver was limited by the photodiode and the tradeoff mentioned above might be improved with respect to an increased gain, i.e., a larger sensitivity. An input noise current spectral density of 8.2 pA $Hz^{-1/2}$ up to 1 GHz caused by shot noise from the base current and thermal noise from the feedback resistor was given. With these values, the photoreceiver sensitivities of −24.3 and −22.8 dBm were estimated for 0.5 and 1 Gb s^{-1}, respectively, for a BER of 10^{-9} and $\lambda = 850$ nm.

For $\lambda = 1.3$ µm and $\lambda = 1.55$ µm, SiGe receivers need external Ge or InGaAs photodiodes. A 155 Mb s^{-1} burst-mode receiver in a 0.8 µm SiGe BiCMOS technology with the optical input power range from −27 to −1 dBm at a BER of 10^{-10} was described in [119]. The external pin photodiode contributes 1.2 pF to the input node capacitance and shows a maximum responsivity of 0.9 A W^{-1}. The power consumption is 500 mW with a power supply voltage of 5 V. The chip consumes an area of 4.3 × 4.9 mm^2 and also includes a comparator, a sample-and-hold circuit and digital-to-analog converters are implemented on the chip.

Using a 0.35 µm SiGe BiCMOS technology, the BM receiver of [6] achieved a sensitivity of −30.2 dBm at 1.25 Gb s^{-1} (PRBS=$2^{15}-1$, BER = 10^{-10}). This high sensitivity is achieved with an avalanche photodiode with a multiplication factor $m = 6$. Therefore, it is not astonishing that the sensitivity is higher than in the other designs, due to the fact that the multiplication factor leads to an advantage of about 7.8 dB compared to a pin photodiode without multiplication of the incoming light. Therefore a sensitivity of 22.4 dBm can be assumed for this amplifier in combination with a pin photodiode. The minimum switching time between loud and soft, with a dynamic range of 21 dB, is given with 25.6 ns for a maximum data rate of 1.25 Gb s^{-1}.

Park [120] introduced an OR in 0.8 μm SiGe heterojunction bipolar transistor (HBT) technology. The TIA is designed in a common-base input configuration depicted in Fig. 8.10. A −3 dB bandwidth of 4.1 GHz was reached with a photodiode capacitance of 0.25 pF and a transimpedance gain of 3.2 kΩ. As optical detector an external InGaAs pin photodiode with a responsivity of 0.8 A W^{-1} was used to detect the 1,550 nm light. At 5 Gb s^{-1} the achieved sensitivity was −19 dBm for a BER of 10^{-10}. The power consumption was 65 mW with VC = 2.5 V and VEE = −2.5 V. This design also offers the possibility of parallel channels with −20 dB crosstalk between adjacent channels.

Another design of Park [121] is quite similar to the design mentioned above, but without a common-base circuit. Again the technology was 0.8 μm SiGe HBT. The TIA is depicted in Fig. 8.11. In this design four parallel channels were realized. Each channel dissipates 10 mW with ±2.5 V supply voltage. The InGaAs pin photodiode with 0.25 pF capacitance and a responsivity of 0.8 A W^{-1} detects the 1,550 nm light. The measured results were a 3.9 GHz bandwidth and a sensitivity of −22 dBm at 5 Gb s^{-1} with a BER of 10^{-12}. The crosstalk between adjacent channels was given with below −25 dBm.

Twelve parallel 10 Gb s^{-1} channels in a 0.35 μm SiGe bipolar production technology were presented in [122]. The chip was built as an array with 250 μm channel pitch which leads to crosstalk problems. The minimum input current swing was $I_{PP} = 20\,\mu A$ with a BER of 10^{-10} which is equal to a sensitivity of −18.4 dBm with the values mentioned at the beginning of the chapter. This measurement was done with emulated photodiodes with a capacitance of 150 fF and 11 disturbing channels. In the measurements with real photodiodes having a capacitance of 250 fF, the eye for the undisturbed channel is still wide

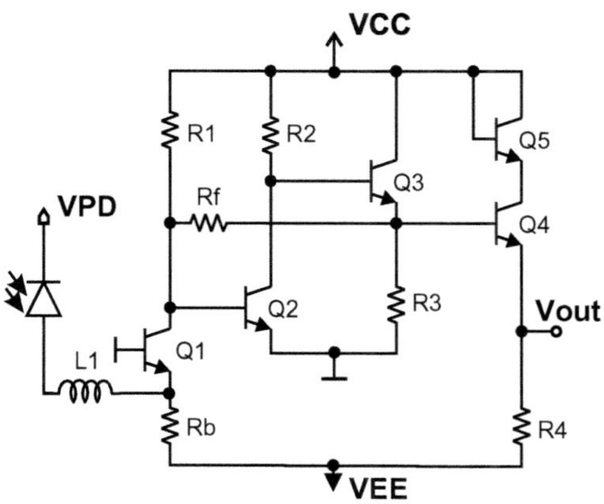

Fig. 8.10. Schematic diagram of single-channel common-base TIA in SiGe HBT technology [120]

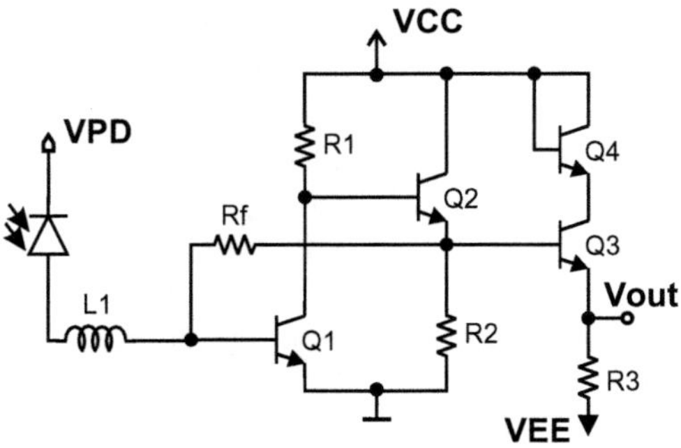

Fig. 8.11. Schematic diagram of single-channel TIA in Si–SiGe HBT technology [121]

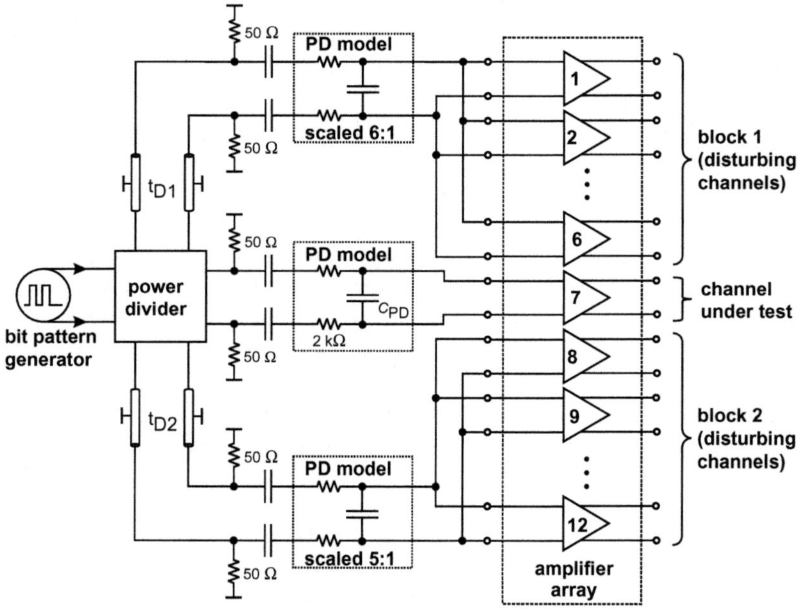

Fig. 8.12. Measurement setup for crosstalk analysis of parallel channels [122]

open, but the opening of the eye with 11 channels disturbing is starting to close. Figure 8.12 shows the measurement setup for crosstalk analysis. The channel under test is channel 7 disturbed by the other channels summarized in two blocks.

In [123] a 10 Gb s^{-1} OR is described with a lateral silicon-on-isolator (SOI) pin photodiode, produced in a 130 nm CMOS technology, wire bonded to a TIA in a 0.18 μm SiGe BiCMOS technology. The photodiode is driven in the avalanche gain mode with a bias voltage of 28 V, and shows an avalanche

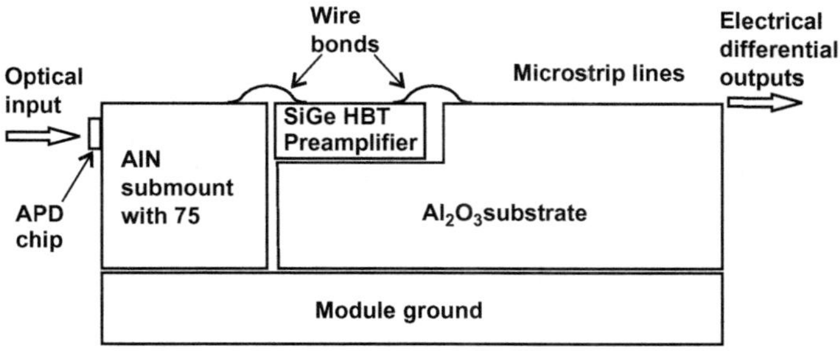

Fig. 8.13. Cross section of OR module (APD and preamplifier) [124]

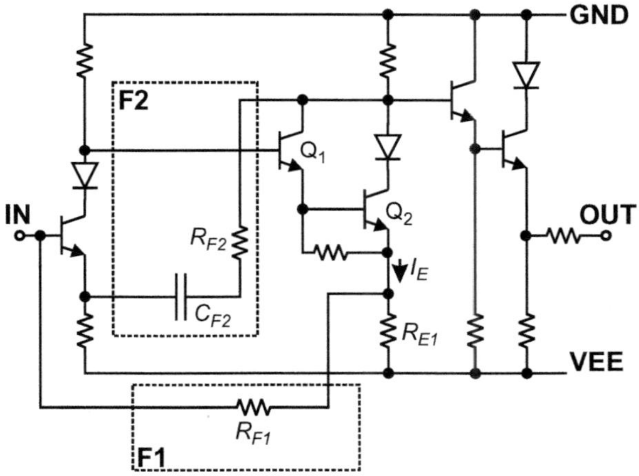

Fig. 8.14. Schematic of a $20\,\mathrm{Gb\,s^{-1}}$ OR in SiGe-base bipolar technology [125]

gain of four which leads to a responsivity of $0.32\,\mathrm{A\,W^{-1}}$ at $\lambda = 850\,\mathrm{nm}$. The measured sensitivity was $-6.9\,\mathrm{dBm}$ at $10\,\mathrm{Gb\,s^{-1}}$ with a BER of 10^{-9}.

Combining an InAlAs avalanche photodiode with an SiGe-HBT preamplifier [124] realized an OR module with a data rate of $10\,\mathrm{Gb\,s^{-1}}$. The parts were mounted together as shown in Fig. 8.13. The achieved sensitivity was $-29.5\,\mathrm{dBm}$ at $10\,\mathrm{Gb\,s^{-1}}$ with a BER of 10^{-9} with 1,550 nm light. The multiplication factor of the APD is 10 which leads also to an advantage in sensitivity of 10 dB compared to a pin photodiode.

The SiGe-base bipolar TIA shown in Fig. 8.14 is part of an OR working at $20\,\mathrm{Gb\,s^{-1}}$ [125]. This design uses an erbium-doped optical fiber amplifier (EDFA) for a high input sensitivity. This EDFA leads to input currents in the mA range. The TIA itself shows two feedback circuits, a current feedback (F1) and a voltage feedback (F2). F1 reduces the input impedance and enables the high bandwidth while F2 minimizes the phase delay in the loop.

A $40\,\mathrm{Gb\,s^{-1}}$ analog front-end for an OR in SiGe HBT technology is described in [126]. With a feedback resistance $R_\mathrm{F} = 750\,\Omega$, a $-3\,\mathrm{dB}$ bandwidth

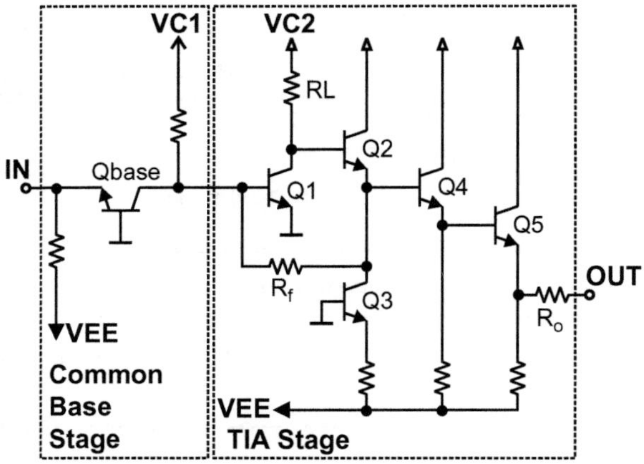

Fig. 8.15. Schematic of a 40 Gb s^{-1} TIA in SiGe HBT [126]

Table 8.2. Summary of sensitivities of state of the art of SiGe designs

reference	process	sensitivity	power cons. (mW)	in. pow. range (dB)
[119]	0.8 µm SiGe BiCMOS	−27 dBm @ 155 Mb s^{-1}	500	26.0
[6]	0.35 µm SiGe BiCMOS	−30.2 dBm @ 1.25 Gb s^{-1} [a]		21
[120]	0.8 µm SiGe HBT	−19 dBm @ 5 Gb s^{-1}	65	
[121]	0.8 µm SiGe HBT	−22 dBm @ 5 Gb s^{-1}	10	
[122]	0.35 µm SiGe bipolar	−18.4 dBm @ 10 Gb s^{-1}	117	
[123]	0.18 µm SiGe BiCMOS	−6.9 dBm @ 10 Gb s^{-1}		
[124]	SiGe HBT	−29.5 dBm @ 10 Gb s^{-1} [b]		

[a] APD, $m = 6$
[b] APD, $m = 10$

of 35.1 GHz is achieved in a common-base TIA configuration (see Fig. 8.15). The influences of the photodiode are taken into account, but they are not specified in more detail. Three power supply voltages are necessary. VC1 is 8.0 V, VC2 equals 5.0 V and VEE equals −5.0 V. The power consumption is 270 mW. An overview of the presented designs is shown in Table 8.2.

8.3 Silicon CMOS Optical Receivers

In [127], a CM OR with capacitive feedback in 0.6 µm CMOS with an average input noise current of 4.5 pA Hz$^{-1/2}$ at a data rate of 622 Mb s^{-1} was presented, see Fig. 8.16. C_1 is sensing the voltage across C_2 and returns a proportional current. This means that C_1 and C_2 replace a feedback resistor R_F when we consider the source of M_2 as an inverting output. Due to the fact that the input photocurrent contains a dc-component a bias network

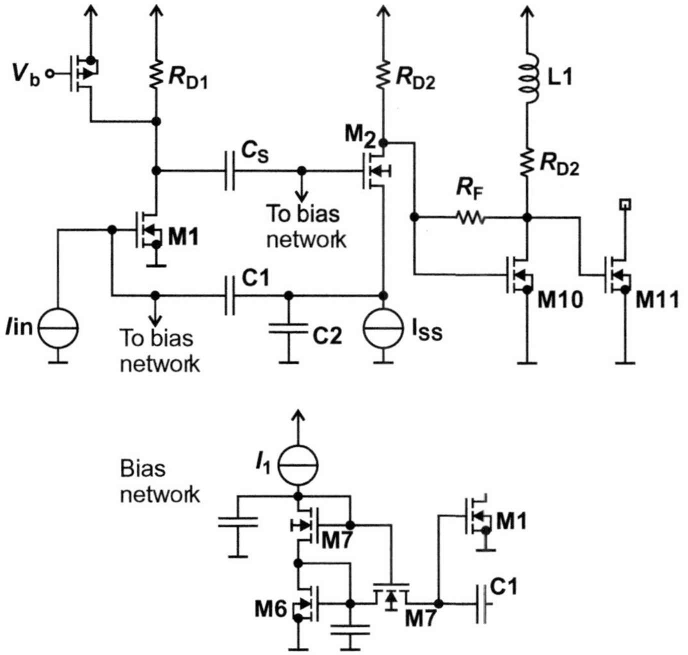

Fig. 8.16. Capacitive feedback OR [127]

consisting mainly of a PMOS current source as shown in Fig. 8.16 is necessary. The capacitive feedback has the advantage that there is no noise source in the feedback path. The average input noise current leads with the photodiode and wavelength data of our own work to a sensitivity of −30.9 dBm at 622 Mb s^{-1}. The power dissipation was 30 mW.

A CM OR with classical inverter structure, depicted in Fig. 8.17, was presented in [128] in a 0.7 μm CMOS technology. The measurements were done with an emulated photodiode and lead to an input current of 10 μA at a data rate of 1 Gb s^{-1}. The photodiode capacitance was given with 0.8 pF and the power consumption of the chip was 100 mW. The sensitivity leads, with the typical values of our own designs, to a sensitivity of −19.2 dBm under the assumption that the given value for the input current is the average input current, and to a sensitivity of −21.8 dBm under the assumption that the given value of 10 μA is the peak value. It was not clearly specified in the paper.

A folded-cascode CM OR was first presented in [98]. The used technology was a 0.35 μm CMOS process and the power consumption of this design was 260 mW at a supply voltage of 2.85 V. With a photodiode responsivity of 0.75 A W^{-1} and a photodiode capacitance of 0.7 pF sensitivities of −22.5 and −27 dBm at 1.25 and 622 Mb s^{-1}, respectively, with PRBS = $2^{23} - 1$ and BER = 10^{-12} were achieved in the circuit simulator.

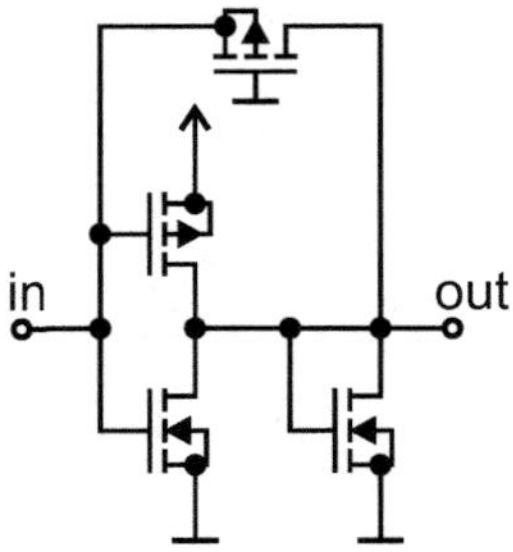

Fig. 8.17. 1 Gb s^{-1} CMOS CM OR [128]

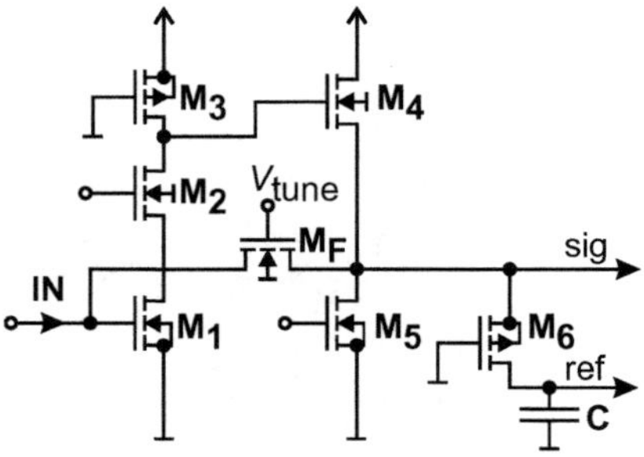

Fig. 8.18. Ethernet-compatible OR in 0.25 μm CMOS technology [129]

Figure 8.18 shows another CM OR in 0.25 μm CMOS technology for 1.25 Gb s^{-1} [129]. The input amplifier is carried out as a cascode stage with the cascode transistor M_2 and a load transistor M_3. This input stage is followed by a source follower stage consisting of M_4 and M_5. The feedback is realized by the transistor M_F. M_6 as an active resistor and C generate the decision threshold for the following differential amplifier. With a photodiode responsivity of 0.5 A W^{-1}, sensitivities of −24 and −14 dBm at 1.25 and 1.5 Gb s^{-1}, respectively, were obtained by on-wafer measurements in [129] with a BER = 10^{-9}. For these measurements the band-limiting element was the postamplifier, not the input node with the photodiode capacitance. The photodiode capacitance was estimated with 110 fF. The reported power consumption of 26 mW is rather low. The maximum achievable data rate of this chip was 2.1 Gb s^{-1} without a mentioned sensitivity.

In [130] a wafer-bonded system is presented with a 0.5 μm CMOS receiver and a GaAs and InGaAs photodiode, respectively. The achieved sensitivities for a GaAs photodiode with a wavelength of 850 nm were −30.1, −27.4, and −18.5 dBm at 622 Mb s^{-1}, 1 Gb s^{-1}, and 1.3 Gb s^{-1}, respectively. For a wavelength of 1,550 nm with the InGaAs photodiode the achieved sensitivities were −31.4 dBm at 622 Mb s^{-1}, −28.0 dBm at 1 Gb s^{-1}, and −22.8 dBm at 1.3 Gb s^{-1}. All sensitivities were given at a BER of 10^{-9} and a PRBS of $2^{15} - 1$. The responsivities of the photodiodes were 0.42 A W^{-1} for the GaAs one and 0.8 A W^{-1} for the InGaAs photodiode, both were produced without antireflecting coating. The dynamic ranges for both types of photodiodes were 22.2 and 22.4 dB at 622 $Mb\,s^{-1}$, at 1 $Gb\,s^{-1}$ the dynamic range was 19.5 and 18 dB for 850 nm and 1,550 nm, respectively. At 1.3 $Gb\,s^{-1}$ the dynamic range went down to 5.5 and 8.0 dB for 850 and 1,550 nm, respectively. The power consumption was given with <30 mW for all circuits. The parasitic capacitances of the photodiodes were given with <10 fF and therefore neglected. Unfortunately the actual capacitances of the photodiodes were not given. The feedback resistance was 5 kΩ.

A TIA with regulated cascode input stage realized in a 0.6 μm CMOS technology is presented in [131]. The regulated cascode (RGC) TIA is depicted in Fig. 8.19. This configuration offers a virtual-ground input impedance due to the fact that the incoming current is converted to a voltage change at the drain of M1 and the local feedback RB and MB reduces the input impedance by the amount of its own voltage gain. The transistors M_2, M_4, and M_5 are connected as source followers. M_3 and R_3 build a common source configuration. The photodiode was emulated by an equivalent circuit on a PCB with a main parasitic capacitance of 1 pF. The measured sensitivity with the equivalent circuit was −20 dBm at 1.25 $Gb\,s^{-1}$ with a BER of 10^{-12}. The power dissipation was 85 mW with a power supply voltage of 5 V.

Burst-mode receivers were first published in the late 1990s. One of the first was [86], shown in Fig. 8.20, which was working at 156 $Mb\,s^{-1}$. It is a three-stage TIA with feed-forward phase compensation of each amplifier stage and a FET-feedback included in each amplifier stage. The FET-feedback means that the output signal of each stage is fed back to the drain of the common-source feedback without interfering with the feed-forward phase compensation. This design enables stable operation over the wide frequency bandwidth. The feedback resistor is carried out switchably for a wide input power range. The measured sensitivity was −35 dBm at 156 $Mb\,s^{-1}$ (PRBS = $2^{23} - 1$, BER = 10^{-10}) with a responsivity of the photodiode of 0.83 $A\,W^{-1}$. The power consumption of this 0.5 μm CMOS receiver was 325 mW with a supply voltage of 3.3 V. The switching between minimum and maximum optical power, when the dynamic range equals 26 dB, is finished in 100 ns.

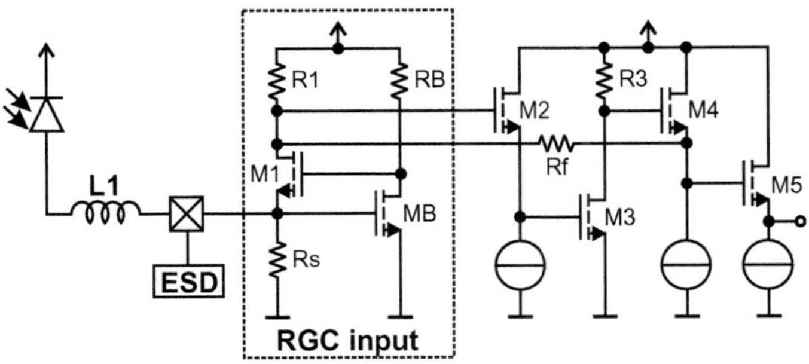

Fig. 8.19. Regulated cascode CMOS TIA [131]

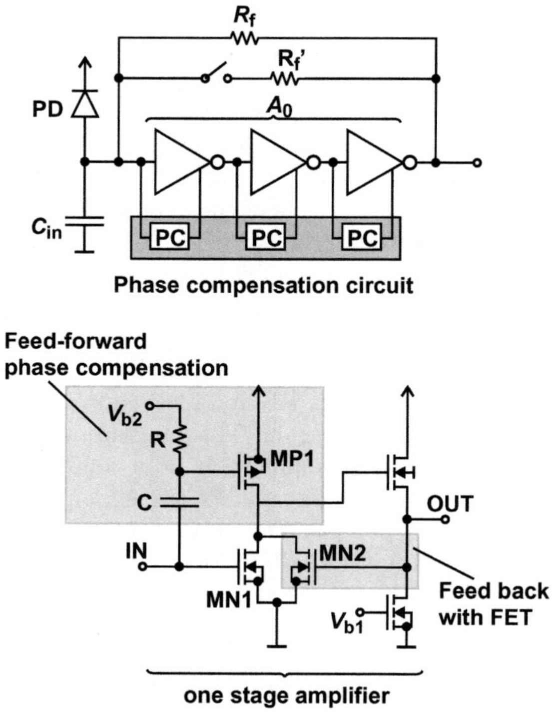

Fig. 8.20. Burst-mode receiver for 156 Mb s^{-1} [86]

Also a burst-mode receiver was presented in [8]. It works at 156 Mb s^{-1} and shows a maximum sensitivity of −39.3 dBm and a dynamic range of 33.3 dB with a responsivity of the photodiode of 0.9 A W^{-1} and a photodiode capacitance of 0.6 pF. The switching is finished after the first bit (time slot of one bit at 156 Mb s^{-1} is about 6 ns). The sensitivities are given at a BER $= 10^{-10}$ and a PRBS $= 2^{23} - 1$. The chip was produced in a 0.25 µm CMOS technology with a supply voltage of 2.5 V. The power dissipation is given with 60–71.4 mW.

In [132] a burst-mode receiver in 0.18 µm CMOS technology is presented with a photodiode responsivity of 0.85 A W^{-1} with a photodiode capacitance of 1 pF. The circuit is depicted in Fig. 8.21. The circuit consists of a cascoded input stage and source followers. The actual TIA is followed by a top- and bottom-level detector which are used for gain control. The gain control varies the feedback resistance value, consisting of R_{FIX} and R_{AGC} in parallel with the feedback transistor M_{FB}, by varying the gate voltage of M_{FB}. The achieved sensitivity was −22 dBm at 1.25 Gb s^{-1} with a PRBS of $2^{23} - 1$ and a BER of 10^{-12} and an input power range of 18.5 dB was measured. The loud/soft switching time was given with 250 ns.

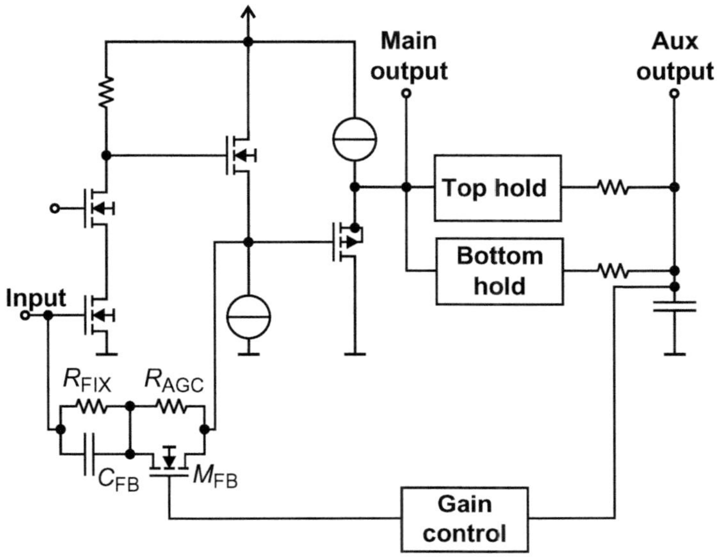

Fig. 8.21. BMR for 1.25 Gb s^{-1} in 0.18 μm CMOS technology [132]

Another fast switching burst-mode receiver was published in [3]. 0.35 μm CMOS was the chosen technology and the *simulated* sensitivity at 1.25 Gb s^{-1} was given with −27 dBm at a BER of 10^{-12}. The switching time for a dynamic range of 21 dB was given with 74 ns. An ac-coupled scheme is used after the preamplifier with a lower −3 dB cut-off frequency of 14.5 MHz. This may lead to problems when long "0" or "1" sequences are sent. The schematic of the preamplifier is equal to that shown in Fig. 8.21, if the top and bottom hold and the gain control is omitted.

In [5] an optical burst-mode receiver is simulated in 0.18 μm CMOS technology. Figure 8.22 shows a block diagram of the circuit. The maximum data rate was 1.25 Gb s^{-1} and the possible sensitivity was given with −29 dBm at this data rate with a photodiode responsivity of 0.85 A W^{-1}. The capacitance of the photodiode was not given. It should be emphasized that this sensitivity is a simulated value and experience has shown that the measured sensitivities normally do not meet the expectations. Power consumption was given with 250 mW at maximum. The dynamic range was given with 21 dB and the switching time was given with 15–40 ns.

A 1.25 Gb s^{-1} burst-mode receiver in 0.25 μm CMOS was presented in [133]. A differential TIA with replica amplifier for the biasing of the second input is used in this design (see Fig. 8.23). The PMOS loads M3 and M4 in the differential amplifier were biased with $V_{GS} = -$AVDD to obtain small channel width and therefore small parasitic node capacitances. Common-mode feedback (CMFB), therefore, is necessary to guarantee a useable operating point in a wide dynamic range. The simulation estimated the photodiode with

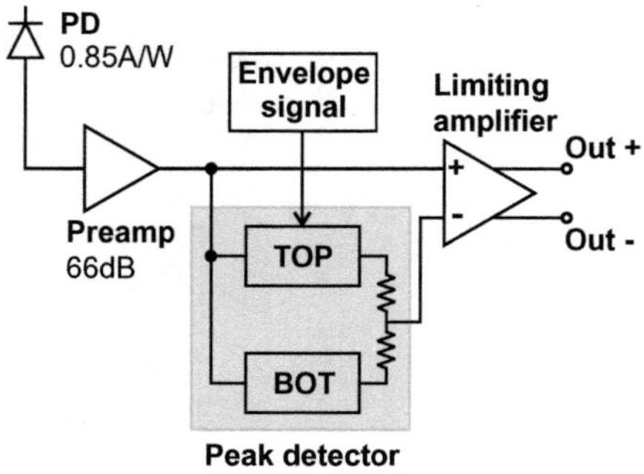

Fig. 8.22. BMR with peak detector for $1.25\,\mathrm{Gb\,s^{-1}}$ [5]

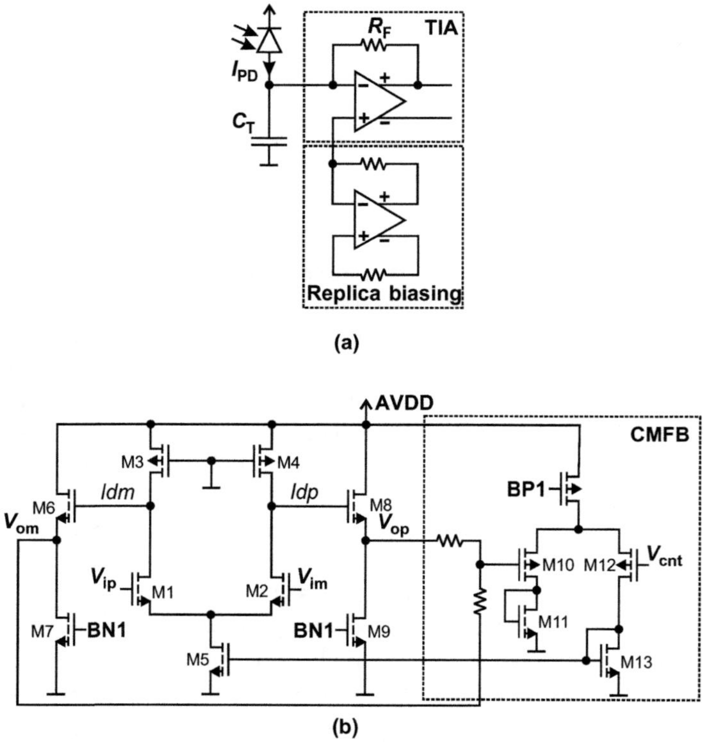

Fig. 8.23. BM OR with automatic level restoration for $1.25\,\mathrm{Gb\,s^{-1}}$: [133] (**a**) block diagram; (**b**) open-loop amplifier

a responsivity of $0.9\,\mathrm{A\,W^{-1}}$ and a photodiode capacitance of 350 fF. This leads to a simulated sensitivity of $-24\,\mathrm{dBm}$ for $\lambda = 1.3\,\mu\mathrm{m}$ at $1.25\,\mathrm{Gb\,s^{-1}}$ with an extinction ratio EX = 5.22 dB. The dynamic range was given with 21 dB and a switching time between minimum and maximum input power was 29 ns.

In [2] a 2.5 Gb s^{-1} CM OR is presented in a 0.35 μm CMOS technology. The input currents were given with 8 and 16 μA for 1.25 and 2.5 Gb s^{-1}, respectively. Sensitivities for $\lambda = 1.3$ μm equal −22.8 and −19.8 dBm at 1.25 and 2.5 Gb s^{-1}, respectively, under the assumption that the given values are peak currents and −20.26 and −17.25 dBm for the two data rates, under the assumption that the given currents are average values. The calculation of the sensitivities in dBm was again done with the characteristic values of the photodiode of our own designs. A BER was not given and the PRBS was $2^{23} - 1$. The power consumption of the chip was 100 mW.

Another CMOS OR front-end for 2.5 Gb s^{-1} was described in [134]. A 0.25 μm CMOS technology was used and the preamplifier reached a transimpedance gain of 250 Ω. It is a single-inverter input stage with an NMOS diode load. The feedback resistance is realized by a transistor. Again the photodiode was emulated and a capacitance of 275 fF was used for the measurements. The equivalent input noise current was given with 9 pA Hz$^{-1/2}$ which equals a sensitivity of −25 dBm at $\lambda = 1.3$ μm for the given values at the beginning of this section.

A fast OR with a maximum bandwidth of 3.5 GHz is presented in [135]. This circuit is designed in a 0.5 μm CMOS technology with an assumed input capacitance of a pin photodiode of 250 fF. The input of this circuit is realized by the transistor M_1 in common-gate configuration (see Fig. 8.24). The output of the common-gate stage of M_1 is fed to a source-follower stage consisting of M_2 and M_4. This is followed by another common-gate configuration including M_5. The drain path of M_5 includes an integrated 5 nH inductor which consumes a large area. This inductor is used to enhance the bandwidth (Fig. 8.24). The cascoded output stage consisting of M_6 and M_7 and R_5 provides additional gain. The simulated power consumption is given with 70 mW.

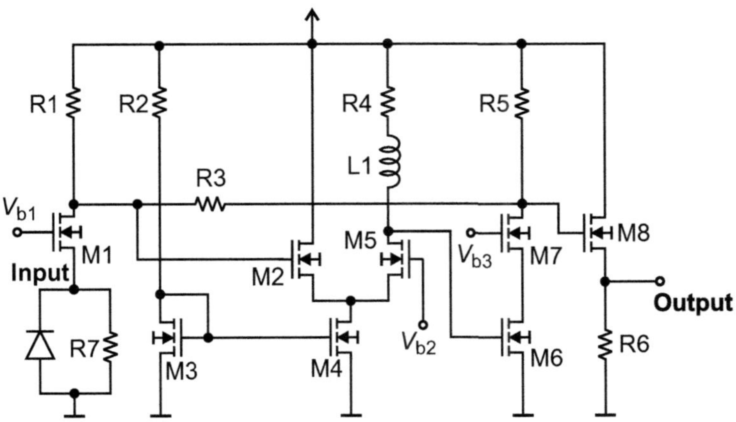

Fig. 8.24. Continuous-mode OR for 3.5 Gb s^{-1} [135]

In [136] a 5 Gb s^{-1} CMOS OR front-end is presented. A 0.18 μm CMOS technology with a supply voltage of 1.8 V was used. The emulated photodiode had a capacitance of 200 fF. Figure 8.25a shows a principle block diagram of the whole circuit. The input voltage is converted by the input test circuit into the differential input current I_{IN}. The next element in the chain is the differential preamplifier which is depicted in detail in Fig. 8.25b. The circuit is designed in common-source configuration, carried out with a cascoded input with the cascode transistors M_c. The constant-k filter at the input is used to minimize the noise. The achieved −3 dB cutoff frequency was 2.6 GHz and the power dissipation of the TIA was 47 mW. The average equivalent input noise current density from 50 MHz to 5 GHz of 13 pA Hz^{-1} equals a sensitivity of −21.1 dBm for $\lambda = 1.3$ μm compared to the other designs.

In 130 nm CMOS technology an integrated silicon OR on silicon-on-isolator (SOI) was introduced by [137]. A pin photodiode was integrated in the CMOS technology to detect the 850 nm light. The presented results, with a reverse voltage at the photodiode of 27 V and therefore operating in the avalanche mode, were +2 dBm at 8 Gb s^{-1}, −10.9 dBm at 5 Gb s^{-1}, −15 dBm

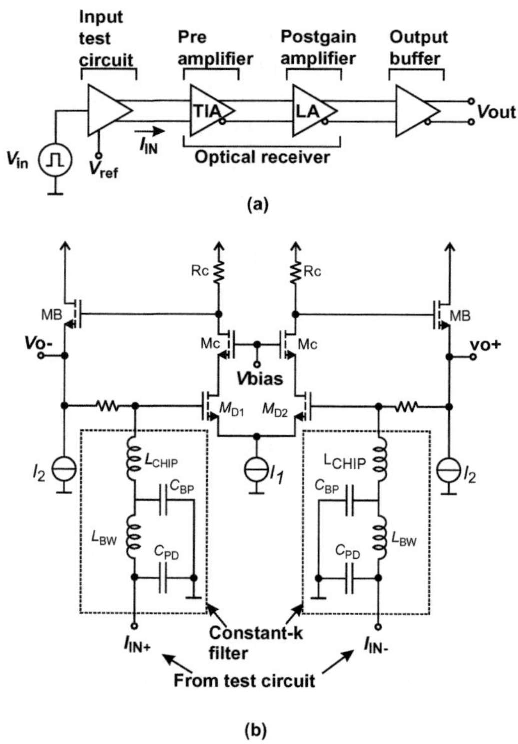

Fig. 8.25. 5 Gb s^{-1} CMOS OR front-end: (**a**) block diagram; (**b**) transimpedance amplifier circuit [136]

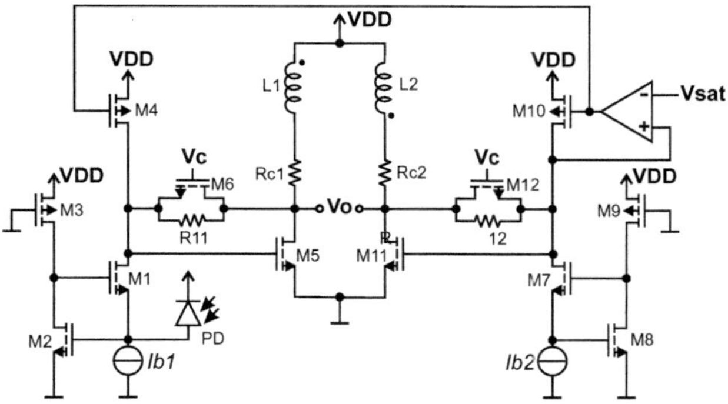

Fig. 8.26. TIA circuit of 10 Gb s^{-1} CMOS OR [138]

at 3.125 Gb s^{-1} and −21 dBm at 2 Gb s^{-1}, respectively, for a BER of 10^{-9} and a PRBS of $2^7 - 1$. Up to 3.125 Gb s^{-1} a single supply voltage of 3 V was possible. The sensitivity at 3.125 Gb s^{-1} was −1 dBm for this case.

An external photodetector Oepic-P5030A with a responsivity of 1 A W^{-1} was used with a 0.18 μm CMOS TIA in [138]. The TIA was designed with a regulated cascode (M1/M2 and M7/M8, see Fig. 8.26) and the achieved −3 dB cut-off frequency was 8 GHz. L1 and L2 boost the bandwidth. The differential amplifier with M5 and M11 is built without a constant current source. The V_{sat} input sets the operating point especially the V_{GS} of M5 and M11. At 10 Gb s^{-1} the sensitivity of the OR for $\lambda = 1.3$ μm was −13 dBm at a BER of 10^{-12} and a PRBS $2^{31} - 1$. The summary of the Si CMOS designs is presented in Table 8.3.

8.4 Summary of Results of State-of-the-Art Optical Receivers

All designs mentioned above with a published sensitivity are summarized in Fig. 8.27. It is difficult to compare the different technologies as well as different used photodiodes and measured and simulated or calculated results. Therefore, we distinguished between Si bipolar and Si BiCMOS processes (Si BJT/BiCMOS), SiGe bipolar and SiGe BiCMOS technologies (SiGe BJT/BiCMOS), and Si CMOS technologies. Furthermore the type of photodiode is extra marked. Designs which modeled the photodiode with electrical circuits for characterization of the design are marked with emulated photodiode (emul. PD). Most important is also to mark the designs with simulated or calculated sensitivities, because experience shows that the simulated or calculated values are normally not met with measurements.

For very high data rates the photodiode capacitance has to be kept small to achieve the necessary bandwidth. On the other hand, this low input-node capacitance offers the advantage that the noise, which is increasing with the

Table 8.3. Summary of sensitivities of state of the art Si CMOS designs

reference	CMOS process (μm)	sensitivity	power cons. (mW)	input power range (dB)
[127]	0.6	−30.9 dBm @ 622 Mb s^{-1} [a]		
[128]	0.7	−21.8 dBm @ 1 Gb s^{-1} [a]	100	
[98]	0.35	−27 dBm @ 622 Mb s^{-1}	260	
[98]	0.35	−22.5 dBm @ 1.25 Gb s^{-1}	260	
[129]	0.25	−24 dBm @ 1.25 Gb s^{-1}	26	
[129]	0.25	−14 dBm @ 1.5 Gb s^{-1}	26	
[130]	0.5	−30.1 dBm @ 622 Mb s^{-1}	<30	22.2
[130]	0.5	−27.4 dBm @ 1 Gb s^{-1}	<30	19.5
[130]	0.5	−18.5 dBm @ 1.3 Gb s^{-1}	<30	5.5
[130]	0.5	−31.4 dBm @ 622 Mb s^{-1}	<30	22.4
[130]	0.5	−28 dBm @ 1 Gb s^{-1}	<30	18
[130]	0.5	−22.8 dBm @ 1.3 Gb s^{-1}	<30	8.0
[131]	0.6	−20 dBm @ 1.25 Gb s^{-1}	85	
[86]	0.5	−35 dBm @ 156 Mb s^{-1}	325	26.0
[8]	0.25	−39.3 dBm @ 156 Mb s^{-1}	71.4	33.3
[132]	0.18	−22 dBm @ 1.25 Gb s^{-1}		18.5
[3]	0.35	−27 dBm @ 1.25 Gb s^{-1} [b]		21
[5]	0.18	−29 dBm @ 1.25 Gb s^{-1} [b]		21
[133]	0.25	−24 dBm @ 1.25 Gb s^{-1}		21
[2]	0.35	−22.8 dBm @ 1.25 Gb s^{-1} [a]		
[2]	0.35	−19.8 dBm @ 2.5 Gb s^{-1} [a]		
[134]	0.25	−25 dBm @ 2.5 Gb s^{-1} [a]		
[136]	0.18	−21.1 dBm @ 5 Gb s^{-1} [a]	47	
[137]	0.13	−21 dBm @ 2 Gb s^{-1} [c]		
[137]	0.13	−15 dBm @ 3.125 Gb s^{-1} [c]		
[137]	0.13	−10.9 dBm @ 5 Gb s^{-1} [c]		
[137]	0.13	+2 dBm @ 8 Gb s^{-1} [c]		

[a] Calculated value out of given input noise value and data of the photodiode described in the beginning of the chapter
[b] Simulated result
[c] Integrated photodiode

squared frequency and squared input-node capacitances, reduces strongly with the low input-node capacitances. Another fundamental advantage concerning noise is an avalanche photodiode (APD) compared to a pin photodiode. Due to the avalanche multiplication factor the sensitivity is increased, but APDs have the disadvantage that high stabilized reverse voltages are necessary.

Details of the comparison of the sensitivities up to 2.5 Gb s^{-1} are shown in Fig. 8.28. Again the distinction among the designs is the same as in Fig. 8.27.

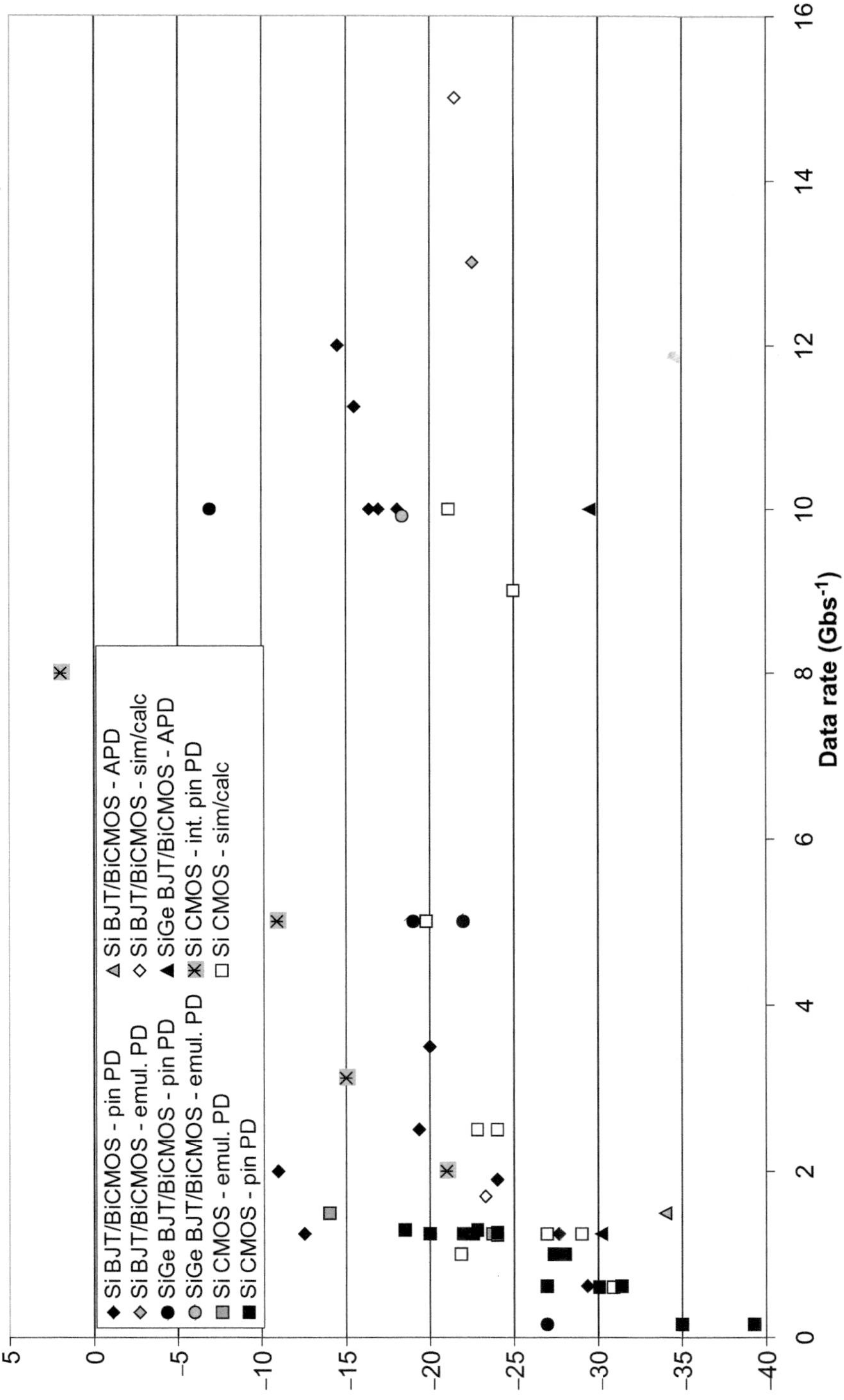

Fig. 8.27. Comparison of sensitivities

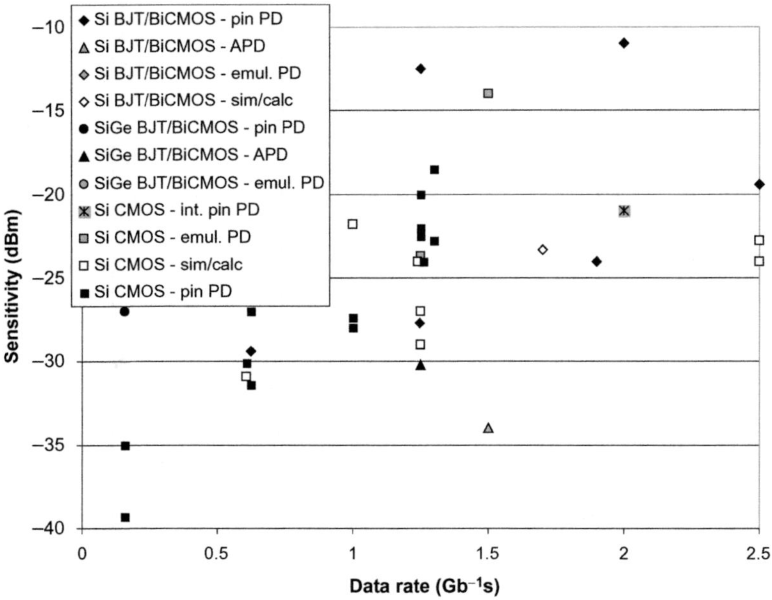

Fig. 8.28. Detail of comparison of sensitivities up to $2.5\,\mathrm{Gb\,s^{-1}}$

Table 8.4. Overview of burst-mode switching times of state of the art

reference	process	maximum data rate	in. pow. range (dB)	switching time (ns)
[86]	0.5 µm CMOS	$156\,\mathrm{Mb\,s^{-1}}$	26.0	100
[8]	0.25 µm CMOS	$156\,\mathrm{Mb\,s^{-1}}$	33.3	6.4
[132]	0.18 µm CMOS	$1.25\,\mathrm{Gb\,s^{-1}}$	18.5	250
[3]	0.35 µm CMOS	$1.25\,\mathrm{Gb\,s^{-1}}$	21	74
[4]	Si bipolar	$1.25\,\mathrm{Gb\,s^{-1}}$	21	75
[107]	Si bipolar	$1.5\,\mathrm{Gb\,s^{-1}}$	15	50
[5]	0.18 µm CMOS	$1.25\,\mathrm{Gb\,s^{-1}}$	21	15–40
[133]	0.25 µm CMOS	$1.25\,\mathrm{Gb\,s^{-1}}$	21	29
[6]	0.35 µm SiGe BiCMOS	$1.25\,\mathrm{Gb\,s^{-1}}$	21	25.6

Burst-mode receivers are also characterized by their switching performance between high and low optical input power. All designs marked as burst-mode receiver and with given switching time are summarized in Table 8.4.

9

Deep-Sub-μm CMOS Circuits

This chapter discusses the simulation environment as well as the transistor models. Mounting and measurement and the characterization setup are described in Sect. 9.2. In Sect. 9.3, the circuits designed in 180 and 120 nm CMOS technology, respectively, are presented together with their results. The chapter ends with a summary and a comparison of our own work to the state of the art depicted in Chap. 8.

9.1 Simulation Environment and Component Models

First, a short overview of the used design environment is given. This is followed by a description of the used photodiode and the resulting model. The top-level design environment will also be discussed briefly. Afterwards in a few lines the transistor models are discussed.

9.1.1 Simulation Environment

The simulations are all done in Cadence. Infineon Technologies provided the models of the components. The schematics are drawn with *Virtuoso Schematic Editor*. The simulations run with *Affirma Analog Artist* which is used with a spectre simulator.

The layouts were drawn in the *Virtuoso Layout Editor*. The design rule check works with the *Assura DRC*, and the layout versus schematic check with the *Assura LVS* tool. A back-annotation tool also exists for Assura, but the back-annotation was made manually, due to the fact that the automatic parasitic extraction did not deliver realistic results.

9.1.2 Photodiode Model

The photodiode used in the designs is an InGaAs/InP-pin-photodiode SRD0014x from Infineon Technologies. It is usable up to 2.5 Gb s^{-1} and

Table 9.1. Typical parameter of SRD0014x pin photodiode

parameter	symbol	value
responsivity @ $\lambda = 1,310\,\mathrm{nm}$	R	$0.85\,\mathrm{A\,W^{-1}}$
max. data rate	DR	$2.5\,\mathrm{Gb\,s^{-1}}$
rise/fall time	$t_\mathrm{r}, t_\mathrm{f}$	$0.2\,\mathrm{ns}$
total capacitance	C	$0.8\,\mathrm{pF}$

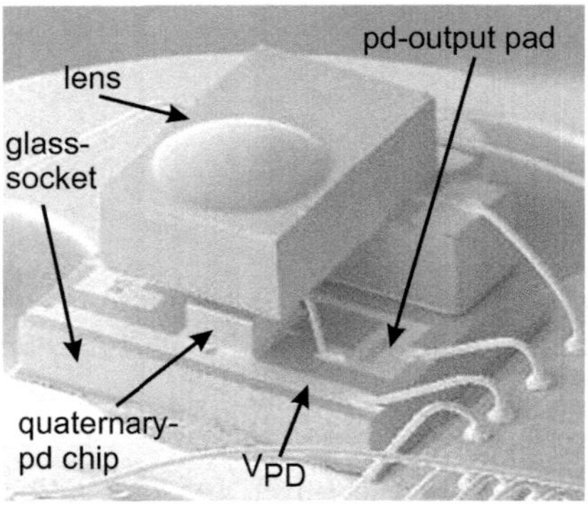

Fig. 9.1. Sketch of photodiode SRD00214x

achieves a responsivity of about $0.85\,\mathrm{A\,W^{-1}}$ at $\lambda = 1{,}310\,\mathrm{nm}$. The main characteristics of the photodiode are summarized in Table 9.1.

The photodiode consists of a quaternary photodiode chip, which is stuck onto a glass socket and covered with a lens. Figure 9.1 displays a sketch of the photodiode. It shows furthermore the pads for the supply voltage and the output of the photodiode, which is connected to the input pad of the TIA chip. The construction of the photodiode with an extra glass-socket necessitates an extra bond wire between the quaternary chip and the glass socket, where the output pad of the photodiode is situated. The supply voltage between socket and photodiode chip is connected via the substrate of the photodiode chip. This leads to the photodiode model shown in Fig. 9.2.

It considers the junction capacitance of the photodiode C_PD the capacitances against ground C_pad of the two pads towards the photodiode output and the supply as well as the bond wires consisting of a series resistor R_b and the inductor L_b, from the actual photodiode to the glass socket and from there to the TIA input. The next following part in the signal chain is the input pad of the TIA, which is considered in the TIA circuit. The bond wires have a diameter of $25\,\mu\mathrm{m}$ and therefore their parameters are estimated with $R_\mathrm{b} = 90\,\mathrm{m}\Omega$ and the inductance with $1\,\mathrm{nH}$ per mm length. This leads to $L_\mathrm{b} = 2\,\mathrm{nH}$ maximum for each wire in the connection from the photodiode to TIA input and therefore this value is chosen for both wires. R_ch represents the channel series resistance of the photodiode.

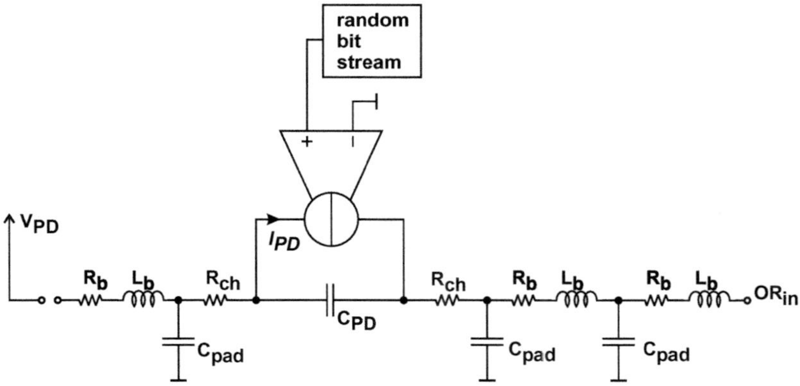

Fig. 9.2. Model for simulation of pin photodiode

The input current of the circuit is simulated by a random bit-stream generator. It produces a pseudorandom bit sequence and offers the possibility to vary the amplitude of the input current to simulate different data bursts.

Top-Level Simulation Environment

Figure 9.3 shows the top-level simulation environment. The burst-mode photodiode model of Fig. 9.2 is connected to the TIA chip input pad. The bond wire between photodiode and TIA is included in the photodiode model and is therefore not visible in this picture. The other bond wires are represented by their resistors and inductors, which are not captioned either. The inductances are estimated between 3 and 5 nH, because the use of equal bond parameters would lead to multiple poles, which may have an unrealistic influence on the performance. The resistors R_o are inserted off-chip for 50 Ω matching of the measurement system.

Due to the fact that the substrate potential defines VSS of the simulation models, the ground outside the chip is denoted as $V_{SS,ext}$ for the external VSS. $V_{DD,ext}$ on the other hand defines the external, off-chip supply voltage. The supply voltage is filtered on chip and therefore the voltage after filtering is defined as VDD and the voltage before filtering is denoted as $V_{DD,ext}$. This ensures the correct simulation of the bond wires and the decreasing of VDD compared to $V_{DD,ext}$ due to the filtering (see Fig. 9.10).

9.1.3 MOSFET Model

The MOSFET model in these designs is a so-called *bsim3v3* model for all designs but the last, where the model was changed to *bsim4*.

The main physical mechanisms the bsim3v3 model accounts for are [139]:

- Short/narrow channel effects on threshold voltage
- Nonuniform doping effects

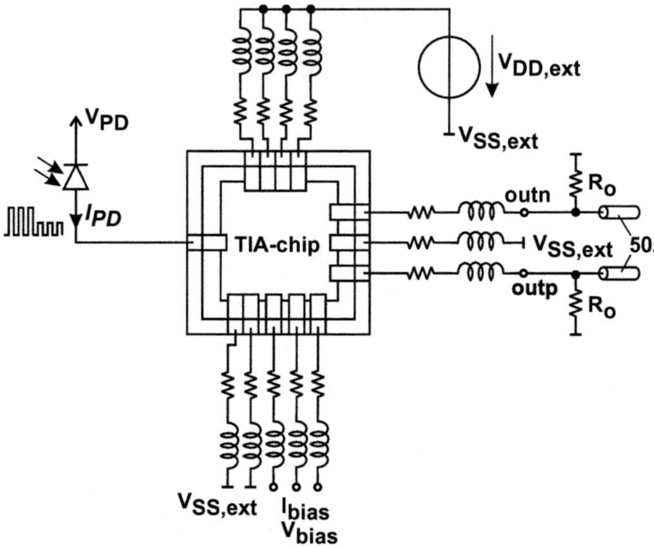

Fig. 9.3. Top-level simulation environment

- Mobility reduction due to vertical field
- Bulk-charge effect
- Carrier velocity saturation
- Drain induced barrier lowering (DIBL)
- Channel length modulation (CLM)
- Substrate current induced body effect (SCBE)
- Parasitic resistance effects
- Quantum mechanic charge thickness model
- Unified flicker noise model

The model is quite complicated and therefore not useful for manual calculations. Figures 9.4 and 9.5 show a principal set of variables of the bsim3v3 model for NMOS and PMOS. These values are not related to any technology; they are typical values for standard technologies [140].

The bsim4 model includes also:

- rf model including the new holistic thermal noise model
- Comprehensive layout-dependent parasitics model
- Quantum mechanical charge thickness model in current/voltage (IV) and capacitnace/voltage (CV)
- Gate-induced drain leakage (GIDL)
- Gate dielectric tunneling current, heavy pocket implant effect

The 120 nm CMOS designs are laid out with so-called rf-cmos transistors. These are transistors with a well-defined layout and a standard width and length. These transistors offer the advantage that the layout is taken into

```
* Predictive Technology Model Beta Version
* 0.13um NMOS SPICE Parametersv (normal one)
*

.model NMOS NMOS
+Level = 49

+Lint = 2.5e-08 Tox = 3.3e-09
+Vth0 = 0.332 Rdsw = 200

+lmin=1.3e-7 lmax=1.3e-7 wmin=1.3e-7 wmax=1.0e-4 Tref=27.0 version =3.1
+Xj= 4.5000000E-08          Nch= 5.6000000E+17
+lln= 1.0000000             lwn= 0.00                  wln= 0.00
+wwn= 1.0000000             ll= 0.00
+lw= 0.00                   lwl= 0.00                  wint= 0.00
+wl= 0.00                   ww= 0.00                   wwl= 0.00
+Mobmod= 1                  binunit= 2                 xl= 0
+xw= 0                      binflag=  0
+Dwg= 0.00                  Dwb= 0.00

+K1= 0.3661500               K2= 0.00
+K3= 0.00                   Dvt0= 8.7500000            Dvt1= 0.7000000
+Dvt2= 5.0000000E-02        Dvt0w= 0.00                Dvt1w= 0.00
+Dvt2w= 0.00                Nlx= 3.5500000E-07         W0= 0.00
+K3b= 0.00                  Ngate= 5.0000000E+20

+Vsat= 1.3500000E+05        Ua= -1.8000000E-09         Ub= 2.2000000E-18
+Uc= -2.9999999E-11         Prwb= 0.00
+Prwg= 0.00                 Wr= 1.0000000              U0= 1.3400000E-02
+A0= 2.1199999              Keta= 4.0000000E-02        A1= 0.00
+A2= 0.9900000              Ags= -0.1000000            B0= 0.00
+B1= 0.00

+Voff= -7.9800000E-02       NFactor= 1.1000000         Cit= 0.00
+Cdsc= 0.00                 Cdscb= 0.00                Cdscd= 0.00
+Eta0= 4.0000000E-02        Etab= 0.00                 Dsub= 0.5200000

+Pclm= 0.1000000            Pdiblc1= 1.2000000E-02     Pdiblc2= 7.5000000E-
03
+Pdiblcb= -1.3500000E-02    Drout= 0.2800000           Pscbe1= 8.6600000E+08
+Pscbe2= 1.0000000E-20      Pvag= -0.2800000           Delta= 1.0100000E-02
+Alpha0= 0.00               Beta0= 30.0000000

+kt1= -0.3400000            kt2= -5.2700000E-02        At= 0.00
+Ute= -1.2300000            Ua1= -8.6300000E-10        Ub1= 2.0000001E-18
+Uc1= 0.00                  Kt1l= 4.0000000E-09        Prt= 0.00

+Cj= 0.0015                 Mj= 0.7175511              Pb= 1.24859
+Cjsw= 2E-10                Mjsw= 0.3706993            Php= 0.7731149
+Cta= 9.290391E-04          Ctp= 7.456211E-04          Pta= 1.527748E-03
+Ptp= 1.56325E-03           JS=2.50E-08                JSW=4.00E-13
+N=1.0                      Xti=3.0                    Cgdo=2.75E-10
+Cgso=2.75E-10              Cgbo=0.0E+00               Capmod= 2
+NQSMOD= 0                  Elm= 5                     Xpart= 1
+Cgsl= 1.1155E-10           Cgdl= 1.1155E-10           Ckappa= 0.8912
+Cf= 1.113e-10              Clc= 5.475E-08             Cle= 6.46
+Dlc= 2E-08                 Dwc= 0                     Vfbcv= -1
```

Fig. 9.4. Variables of bsim3 model for NMOS in 130 nm technology [140]

```
* Predictive Technology Model Beta Version
* 0.13um PMOS SPICE Parametersv (normal one)
*

.model PMOS PMOS
+Level = 49

+Lint = 2.e-08 Tox = 3.3e-09
+Vth0 = -0.3499 Rdsw = 400

+lmin=1.3e-7 lmax=1.3e-7 wmin=1.3e-7 wmax=1.0e-4 Tref=27.0 version =3.1
+Xj= 4.5000000E-08         Nch= 6.8500000E+18
+lln= 0.00                 lwn= 0.00                 wln= 0.00
+wwn= 0.00                 ll= 0.00
+lw= 0.00                  lwl= 0.00                 wint= 0.00
+wl= 0.00                  ww= 0.00                  wwl= 0.00
+Mobmod=  1                binunit= 2                xl=  0
+xw=  0                    binflag=  0
+Dwg= 0.00                 Dwb= 0.00

+K1= 0.4087000             K2= 0.00
+K3= 0.00                  Dvt0= 5.0000000           Dvt1= 0.2600000
+Dvt2= -1.0000000E-02      Dvt0w= 0.00               Dvt1w= 0.00
+Dvt2w= 0.00               Nlx= 1.6500000E-07        W0= 0.00
+K3b= 0.00                 Ngate= 5.0000000E+20

+Vsat= 1.0500000E+05       Ua= -1.4000000E-09        Ub= 1.9499999E-18
+Uc= -2.9999999E-11        Prwb= 0.00
+Prwg= 0.00                Wr= 1.0000000             U0= 5.2000000E-03
+A0= 2.1199999             Keta= 3.0300001E-02       A1= 0.00
+A2= 0.4000000             Ags= 0.1000000            B0= 0.00
+B1= 0.00

+Voff= -9.10000000E-02     NFactor= 0.1250000        Cit= 2.7999999E-03
+Cdsc= 0.00                Cdscb= 0.00               Cdscd= 0.00
+Eta0= 80.0000000          Etab= 0.00                Dsub= 1.8500000

+Pclm= 2.5000000           Pdiblc1= 4.8000000E-02    Pdiblc2= 5.0000000E-
05
+Pdiblcb= 0.1432509        Drout= 9.0000000E-02      Pscbe1= 1.0000000E-20
+Pscbe2= 1.0000000E-20     Pvag= -6.0000000E-02      Delta= 1.0100000E-02
+Alpha0= 0.00              Beta0= 30.0000000

+kt1= -0.3400000           kt2= -5.2700000E-02       At= 0.00
+Ute= -1.2300000           Ua1= -8.6300000E-10       Ub1= 2.0000001E-18
+Uc1= 0.00                 Kt1l= 4.0000000E-09       Prt= 0.00

+Cj= 0.0015                Mj= 0.7175511             Pb= 1.24859
+Cjsw= 2E-10               Mjsw= 0.3706993           Php= 0.7731149
+Cta= 9.290391E-04         Ctp= 7.456211E-04         Pta= 1.527748E-03
+Ptp= 1.56325E-03          JS=2.50E-08              JSW=4.00E-13
+N=1.0                     Xti=3.0                  Cgdo=2.75E-10
+Cgso=2.75E-10             Cgbo=0.0E+00             Capmod= 2
+NQSMOD= 0                 Elm= 5                   Xpart= 1
+Cgsl= 1.1155E-10          Cgdl= 1.1155E-10         Ckappa= 0.8912
+Cf= 1.113e-10             Clc= 5.475E-08           Cle= 6.46
+Dlc= 2E-08                Dwc= 0                   Vfbcv= -1
```

Fig. 9.5. Variables of bsim3 model for PMOS in 130 nm technology [140]

account in the transistor model, based on the bsim models. To generate larger widths the transistor cells have to be placed in parallel.

9.2 Characterization Setup

For the characterization the TIA chip is mounted on a printed circuit board (PCB). It is glued with an electrical and thermal highly conductive paste, which mainly contains silver, onto the PCB. The contacts for biasing, supply voltages and output voltage are bonded to gold-coated pads on the PCB. The dc voltages (biasing and supply voltage) are generated on separate boards and connected via simple connectors (see Fig. 9.6 on top and bottom). They are blocked again by capacitors next to the bond pad at the PCB. The differential output signals are fed into 50 Ω microstrip lines on the PCB and matched with an external SMD resistor. The microstrip lines are fed in the following via SMA-connectors into the measurement system. An example of a board for the characterization is shown in Fig. 9.6. The photodiode has its own supply voltage, which is provided separately from the supply voltage of the TIA chip, on the left-hand side in Fig. 9.6.

The external photodiode is mounted next to the TIA chip on the laminate. They are connected together with a bond wire. To keep the parasitics of the bond wire as small as possible the photodiode is mounted as near as possible to the input pad of the TIA chip. The bond wire directly connects the output of the photodiode with the input of the TIA chip (see Fig. 9.7).

Fig. 9.6. Printed circuit board for characterization

Fig. 9.7. Mounting detail of TIA chip with photodiode

The high capacitance of the photodiode and the bond-pad capacitance cause a rather high input-node capacitance. It results from the capacitance of the photodiode, the bond-pad capacitance and the input capacitances of the input transistors of the TIA. The overall input-node capacitance rises to about 2 pF with a photodiode voltage V_{PD} of 5 V.

The measurement setup is shown in Fig. 7.3. The chip is driven with a data signal at a specified bit rate and an optical input power. The PRBS signal generated by a bit-pattern generator is provided to a laser and the clock is fed into the communication signal analyzer. The generated optical PRBS signal is fed via the photodiode into the TIA chip and the output voltage is measured. For specified BERs at certain data rates the necessary average optical input powers are measured with a calibrated optical power meter.

9.3 Designs and Properties of Optical Receivers

Corresponding to the demands on the optical receivers concerning noise and bandwidth, the following designs are optimized in these directions. Technology was defined as deep-sub-micron CMOS. First a 180 nm CMOS technology was available, the other designs were made in a 120 nm CMOS technology.

The receivers described in the following section, except one design, use a differential structure to decrease the influence of substrate noise from the digital part in a SoC. Additional advantages are the reduction of temperature effects and of effects caused by process tolerances and the biasing is easier. The actual receiver consists of a TIA, a dummy TIA and a differential stage.

The additional noise due to the differential structure is suppressed by a RC-filter in the dummy TIA. A driver circuit for characterization had to be implemented to drive the 50 Ω signal analyzer input (see Fig. 9.8). It consists of two differential stages, followed by a source follower for level shift and to take care of the load of the following stage and an also differential output

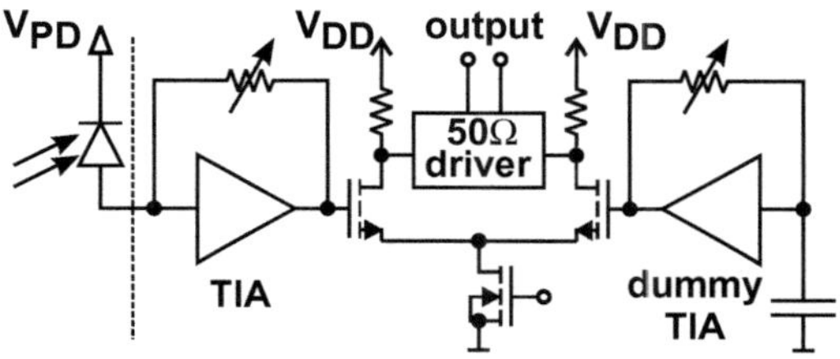

Fig. 9.8. Basic block diagram of differential TIA design

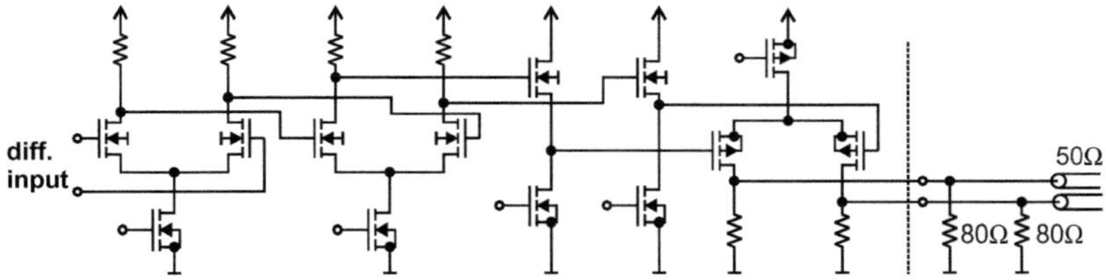

Fig. 9.9. 50 Ω driver, basic schematic

stage to drive the 50 Ω load. The circuit shown in Fig. 9.9 is only a principle circuit. The detailed output drivers for the individual designs slightly vary in the dimensions of the components and the number of level shifters differs as needed for a correct function.

A complete schematic is shown in Fig. 9.10 considering for example the three-inverter TIA. Beside the 50 Ω driver the filtering of the supply voltage and bias voltages is shown as well as the internal biasing of the output driver via current mirrors. In the following the measured output voltage will be called output voltage of the *whole chip* and the output node of the preamplifier, the TIA, only will be called *TIA-output.*

All pads, except the TIA-input pad, are standard pads provided by the foundry as pad libraries. They consist of an ESD-structure and a bonding spot where the actual contact is made. The ESD-structures consist of diodes against plus and ground and a series resistance. The series resistance for an input pad is rather high and therefore it is useful to use output pads with low series resistance for the input of fast switching bias voltages. The bonding spots are carried out as aluminum spots for the actual contact and to guarantee mechanical stability during the bonding floating structures of all metal layers are placed below the aluminum spot.

The TIA-input pad consists of the aluminum bonding spot only; the ESD protection as well as the mechanical stabilization are removed to save capacitance and resistance. The final optical receiver includes the photodiode and the CMOS receiver chip and therefore the lack of ESD protection is no problem for the handling of the complete system.

In the following layout plots of the designs are presented, because the microphotograph of the chip shows no details due to the passivation layer and

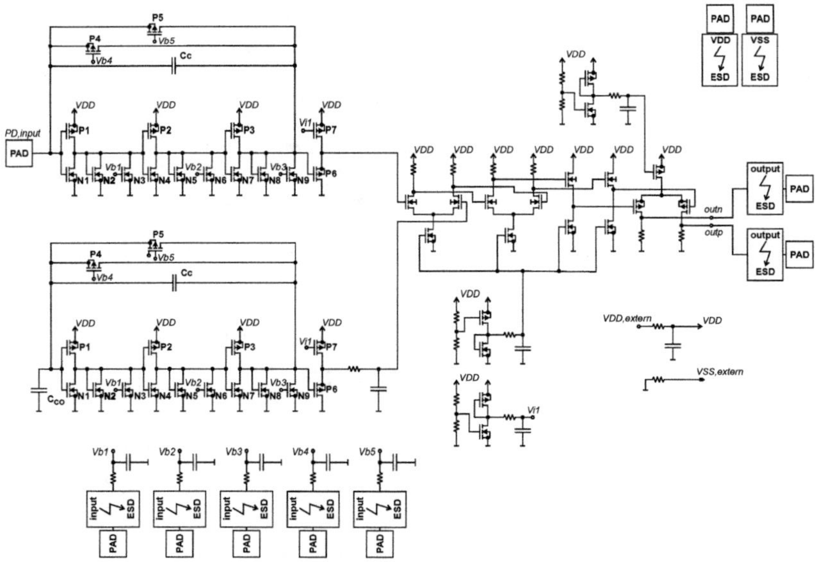

Fig. 9.10. Example of complete schematic (three-inverter amplifier)

the planarization of the metal layers in the deep-sub-micron technology. The only detailed microphotograph presented shows the folded-cascode TIA. This is possible, because the passivation of this chip was removed before.

9.3.1 Folded-Cascode Transimpedance Amplifier

This circuit was designed in a standard digital 180 nm CMOS technology with a power supply voltage of 1.8 V. The design was optimized concerning noise. For high sensitivities the input node capacitance of the TIA must be small, see Sect. 7.3.1. This input node capacitance consists of the capacitance of the photodiode, the input bond-pad capacitance, the feedback capacitance of the TIA and the input capacitance of the input transistor. In common TIA stages like those presented in [141] the Miller capacitance of the input transistor causes an additional capacitance at the input node. In a folded-cascode circuit the Miller capacitance of the input transistor is reduced and therefore the input node capacitance is minimized. The advantage of a folded cascode over a normal cascode is that the drain–source voltage of each transistor is higher. Therefore the width-to-length ratio of each transistor can be kept small and the parasitic capacitance of each transistor will be small too. Due to the fact that the provided supply voltage is 1.8 V it is important that this circuit works with a low number of transistors between VDD and VSS [102].

Figure 9.11 shows the folded-cascode circuit. The input transistor N1 is large to get high gain and a good noise performance, the transistor P1 is the cascode stage, giving low impedance for the drain of N1 and reducing the

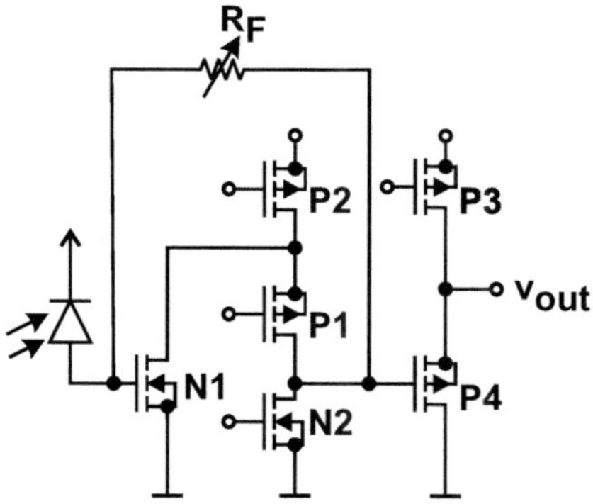

Fig. 9.11. Schematic of the folded-cascode TIA

effective drain-gate capacitance (Miller capacitance of N1). The transistors P2 and N2 are current sources. The output node is of high impedance, therefore a source follower (P3, P4) is used to take care of the load.

The TIA chip contains the TIA itself and a circuit necessary for characterization. To minimize the effects of substrate noise a pseudodifferential design is used. Figure 9.12 shows a microphotograph of the actual TIA chip. This test chip has a total over-all area of 1.014 mm^2. The two TIAs and the differential amplifier, the so-called active area, are located in the T-shaped white frame in Fig. 9.12 and occupy an area of 0.0563 mm^2. When we add the area of the input bond pad, we obtain 0.0703 mm^2. The total area of the folded-cascode differential front-end including blocking capacitors and current mirrors for biasing is about 0.33 mm^2 [142].

Figure 9.13 shows the eye diagram at a data rate of 622 $Mb\,s^{-1}$ for an optical input power of −28.7 dBm in the most sensitive setting ($R_F = 1.6\,k\Omega$) measured with a laser having the extinction ratio of 3. A Q-factor of 6.3 was measured which means a BER of 10^{-10}. The sensitivity of the TIA, due to the power-penalty correction for an extinction ratio of 10 (see Sect. 7.1.3), therefore, is −30.7 dBm. Figure 9.14 shows the eye diagram for a data rate of 1.25 $Gb\,s^{-1}$ in the medium-gain setting ($R_F = 380\,\Omega$). The detected sensitivity is −23.7 dBm at a BER of 10^{-10}. The photodiode was biased at +5 V and VDD was 1.8 V. The current consumption of the whole test chip was 54 mA, whereby the active area consumed 27 mA from the 1.8 V supply.

The folded-cascode TIA for burst-mode applications has a sensitivity of −30.7 dBm at 622 $Mb\,s^{-1}$ and −23.7 dBm at 1.25 $Gb\,s^{-1}$ at a BER of 10^{-10}. Due to the fact that the chip was processed with more metal layers than planned, the input pad was not connected to the inner circuit due to an oxide layer between the actual aluminum pad layer and the following metal layer. Therefore this connection had to be done afterwards. This preparation leads to a series ohmic resistance of about 50 Ω between the input pad and the actual TIA input and therefore the simulated bandwidth could not be achieved, and

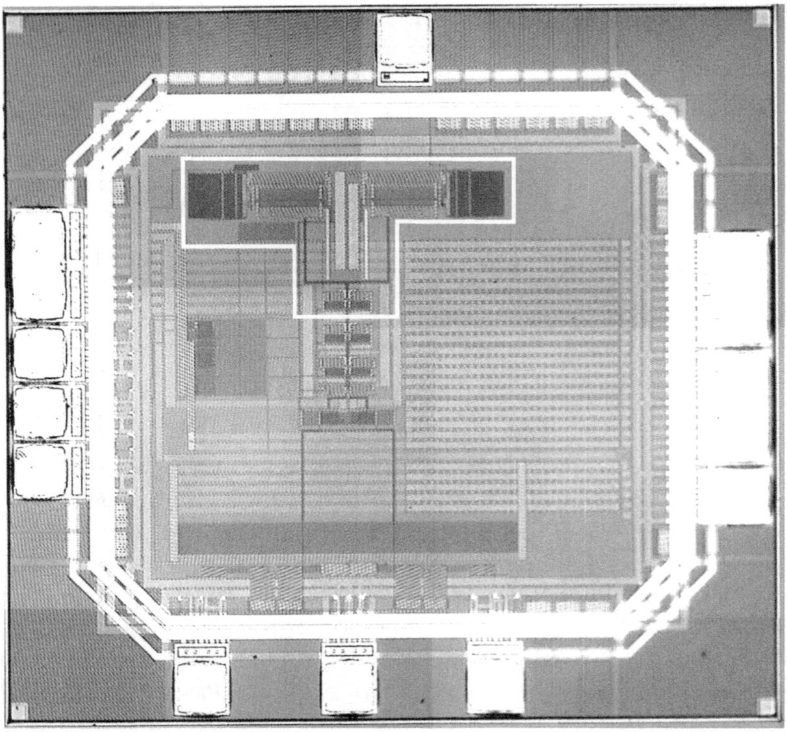

Fig. 9.12. Microphotograph of folded-cascode TIA chip with white marked active area

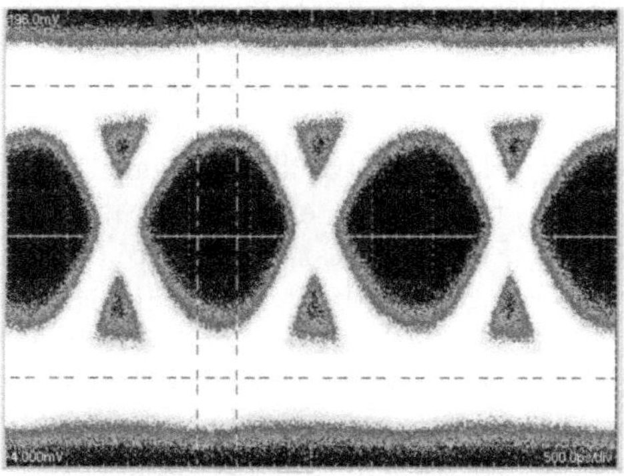

Fig. 9.13. Eye diagram of the folded-cascode TIA at 622 Mb s^{-1} and for −28.7 dBm optical power ($\lambda = 1{,}310$ nm, PRBS $= 2^{31} - 1$, EX $= 3$)

the measured sensitivities were behind the simulated ones, due to the lower bandwidth and the additional noise of the series resistance.

This design is not at its limits; we expect better results of a redesign especially at 1.25 Gb s^{-1}. The results are summarized in Table 9.2.

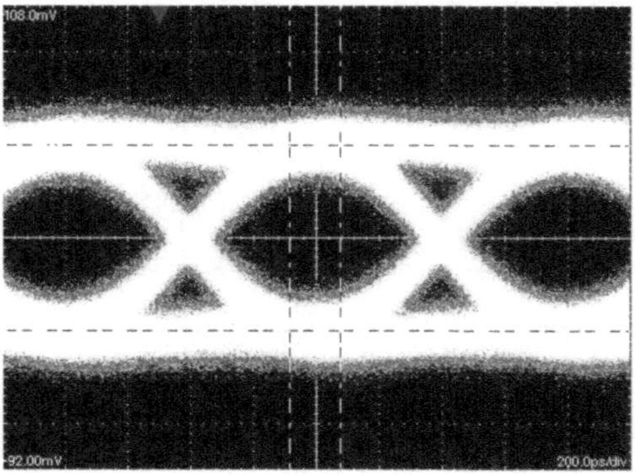

Fig. 9.14. Eye diagram of the folded-cascode TIA at 1.25 $\mathrm{Gb\,s^{-1}}$ and for −21.7 dBm optical power (λ=1,310 nm, PRBS=$2^{31}-1$, EX=3)

Table 9.2. Summarized results of folded-cascode TIA (EX = 10, λ = 1,310 nm, PRBS = $2^{31}-1$)

bit rate	sensitivity (dBm)	BER
1.25 $\mathrm{Gb\,s^{-1}}$	−23.7	10^{-10}
622 $\mathrm{Mb\,s^{-1}}$	−30.7	10^{-10}

9.3.2 Three-Inverter Transimpedance Amplifier

For this design a 120 nm CMOS technology was available. The challenge was to achieve a high open-loop gain for high bandwidth and good noise performance. The power supply voltage is 1.5 V. This very low power supply voltage leads to a low gain for a one-stage amplifier (about 10) in this technology. To achieve a high open-loop gain, an inverter structure was chosen. Inverters can be easily combined to a three-stage system, compared to the folded-cascode amplifier of the design before, which is necessary to achieve the high open-loop gain, to achieve a high bandwidth and a good noise performance.

Figure 9.15 shows the schematic of the three-inverter TIA. The TIA itself is designed as a three-stage amplifier with variable feedback for a wide input-current range. The three stages are necessary to produce a maximum open-loop gain $A_0 = 440$ of the TIA to enable high transimpedance R_{F} despite a high bandwidth $f_{-3\,\mathrm{dB}}$ and high data rates according to (9.1) with the high input-node capacitance C_{T} due to the external photodiode.

The amplifier of the TIA consists of three inverter stages (P1/N1, P2/N4, P3/N7) each with constant diode load (N2, N5, N8) and variable (N3, N6, N9) load. The feedback is designed as transistors (P4 and P5) to form variable resistance values plus a capacitance for compensation (C_{F}). P6/P7 build a source follower to take care of the load. The sensitivity essentially depends on

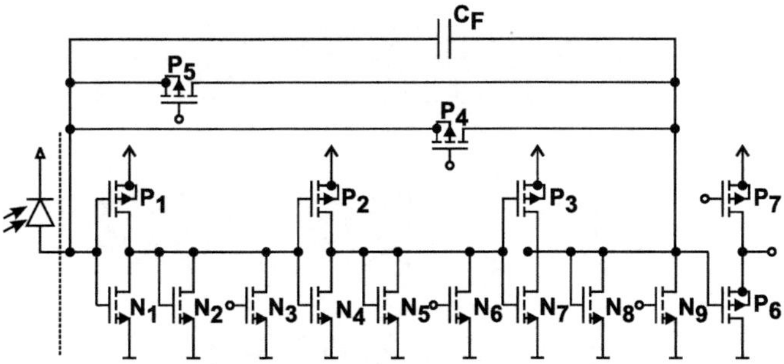

Fig. 9.15. Schematic of three-inverter TIA

the input-node capacitance, the feedback resistor and the transconductance of the input transistors of the amplifier (compare Sect. 7.4.2). To obtain a high sensitivity despite the high input-node capacitance, the transconductance of the input transistors and therefore the current through P1/N1 has to be maximized for minimum noise. The inverter structure shows a better noise behavior than a simple amplifier consisting of a transistor with resistive load, because the noise of the amplifier, which mainly consists of the thermal noise of the input transistors, is divided by the sum of the transconductances of the input transistors in case of the inverter, where the noise of the simple amplifier, consisting of the thermal noise of the transistor and the resistor noise, is only divided by the transconductance of the single transistor. For minimum optical input power and thus minimum input current, which needs the highest transimpedance, only P4 is active in the feedback, and therefore the gain of the amplifier must be maximum for high data rates. To achieve the high gain, the variable loads of the inverters are switched off, by setting the gate voltages of the transistors N3, N6, N9 to zero.

Figure 9.16 shows the simulated ac frequency response for the output of the whole chip and the output node of the TIA itself. A bandwidth of 950 MHz at the chip output node is achieved. The TIA itself shows a bandwidth of 1.4 GHz.

For higher optical input power and therefore higher input currents the transimpedance has to be lower to prevent overdrive. Hence both feedback transistors can be active and to guarantee stability and an approximately constant bandwidth, the variable loads (N3, N6, N9) are activated to lower the gain to $A_0 = 1.3$ at minimum. The resulting wide dynamic range of 340 for A avoids the problem of switching C_F. The resistance of the variable loads can be controlled by their gate voltages. Figure 9.17 shows the layout plot of the chip [143].

In Fig. 9.18 the eye diagram for 1.25 Gb s^{-1} and an average optical input power of -28.2 dBm and a BER of 10^{-10} with a PRBS of $2^{31}-1$ is shown.

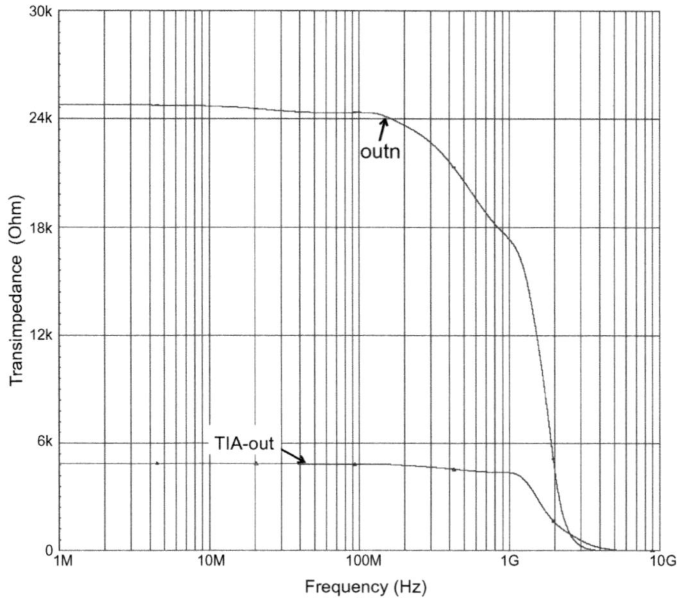

Fig. 9.16. Simulated ac analysis of three-inverter TIA

A sensitivity of −29.9 dBm at a data rate of 622 Mb s^{-1} (see Fig. 9.19) is achieved with the same BER and PRBS. For a BER of 10^{-9} and a PRBS of $2^{31} - 1$ the sensitivity increases to −30.2 dBm at 622 Mb s^{-1} and −28.3 dBm at 1.25 Gb s^{-1}.

The active chip area of the TIA, dummy TIA and the first differential stage consumes 0.1 mm^2. The whole chip area is 1.24 mm^2. Again, the big difference between these two values results from on-chip blocking capacitors and the 50 Ω output driver circuit. The power consumption of the circuit depends on the gain setting of the TIA. For highest sensitivity, the chip consumes 154.5 mW with a power supply voltage of 1.5 V. The actual receiver consumes about 63% of this amount. The rest is dissipated in the driver, which is not necessary in an SoC. The power consumption of the first amplifier stage is about 20 mW. For lowest sensitivity the power dissipation of the chip increases to 210 mW due to the additional loads N3, N6, and N9. The maximum average optical input power is −8.9 dBm at 1.25 Gb s^{-1} and PRBS of $2^{31} - 1$. Table 9.3 summarizes the measured results.

The achieved sensitivities were quite good, but the design has the disadvantage that the power consumption is increasing for decreasing transimpedance gain, due to the opening of the additional loads.

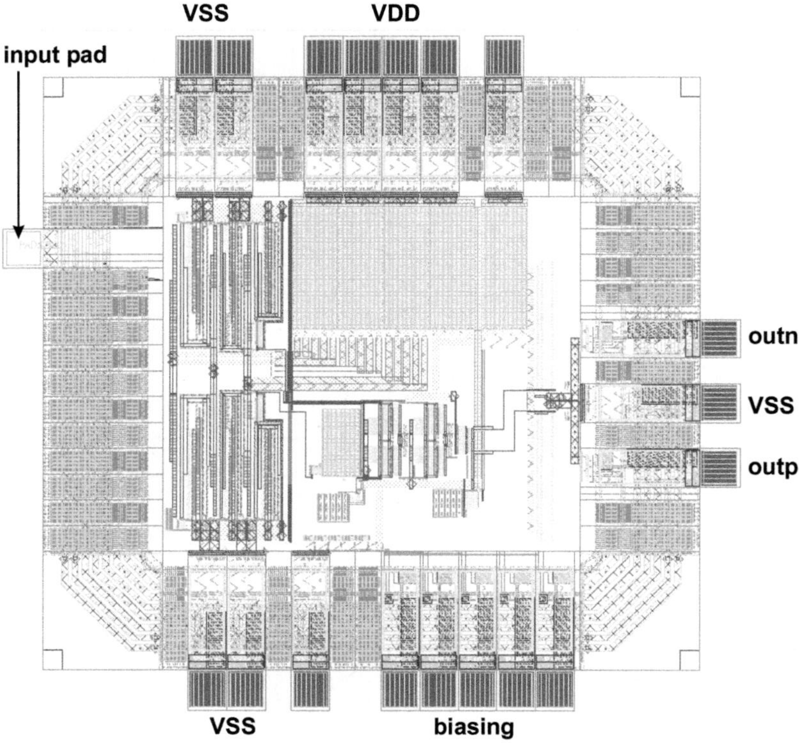

Fig. 9.17. Layout plot of three-inverter TIA

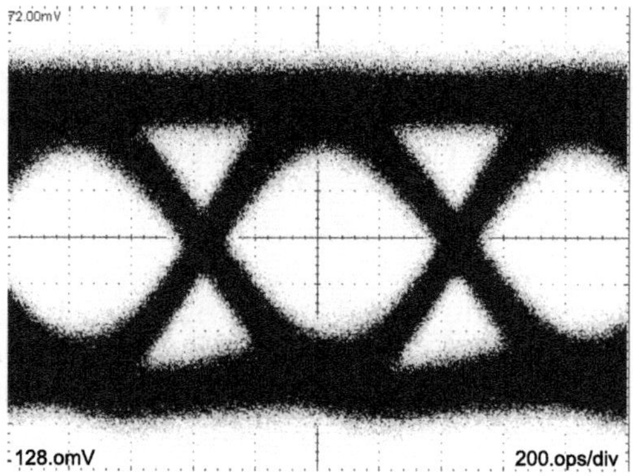

Fig. 9.18. Eye diagram of three-inverter TIA at 1.25 Gb s^{-1} for an average optical input power of -28.2 dBm (BER $= 10^{-10}$, EX $= 10$, $\lambda = 1310$ nm, PRBS $= 2^{31} - 1$)

9.3.3 Three-Stage Transimpedance Amplifier

A 120 nm CMOS technology was used for the next design again. The attention was focused on the power consumption this time. The design was carried out

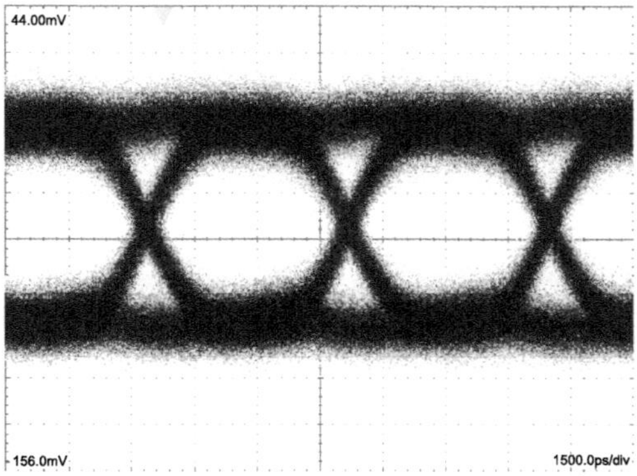

Fig. 9.19. Eye diagram of three-inverter TIA at 622 Mb s^{-1} for an average optical input power of −29.9 dBm (BER = 10^{-10}, EX = 10, λ = 1,310 nm, PRBS = $2^{31} - 1$)

Table 9.3. Summarized results of three-inverter TIA (EX = 10, $\lambda = 1{,}310$ nm, PRBS = $2^{31} - 1$)

bit rate	sensitivity (dBm)	BER
1.25 Gb s^{-1}	−27.5	10^{-12}
622 Mb s^{-1}	−28.5	10^{-12}
1.25 Gb s^{-1}	−28.2	10^{-10}
622 Mb s^{-1}	−29.9	10^{-10}
1.25 Gb s^{-1}	−28.3	10^{-9}
622 Mb s^{-1}	−30.2	10^{-9}

to switch off gain-contributing stages in the amplifier to reduce the open-loop gain. This offers the advantage to reduce the power consumption for lower transimpedance gain [144].

The TIA itself is designed as a three-stage amplifier with variable feedback for a wide input-current range. The three stages result in a high open-loop gain $A_0 = 37.16$ dB and are necessary to enable the high transimpedance of 3 kΩ at high data rates despite the large input-node capacitance.

Figure 9.20 shows the schematic of the TIA. The transistors P1 to P4 and N1 to N4 build the three amplifier stages. The transistors P2 and N2 are used as load for the first inverter-stage P1/N1. The transconductance of P1 $g_{\mathrm{m,P1}} = 47.5$ mS and $g_{\mathrm{m,N1}} = 89$ mS for N1. P2 is a constant diode load and N2 is adjustable to lower the gain in case of instability and therefore guarantee stability. The inverter structure was chosen to minimize the input noise current of the amplifier. P3, N3 show a total transconductance $g_{\mathrm{m,P3}} + g_{\mathrm{m,N3}} = 90$ mS and P4, N4, with $g_{\mathrm{m,P4}} + g_{\mathrm{m,N4}} = 51$ mS, form the amplifier stages two and three, respectively. The p-MOSFETs were used as amplifying transistors, because the gate voltage of P3 was easier to adjust for the p-MOSFET. In the used 120 nm process the p-MOSFETs are not really slower

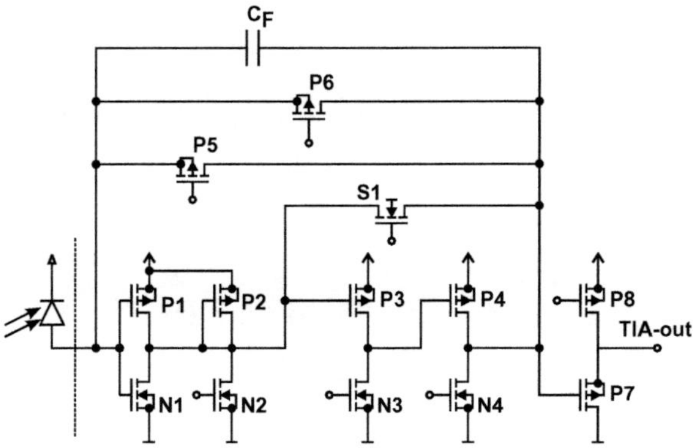

Fig. 9.20. Schematic of three-stage TIA

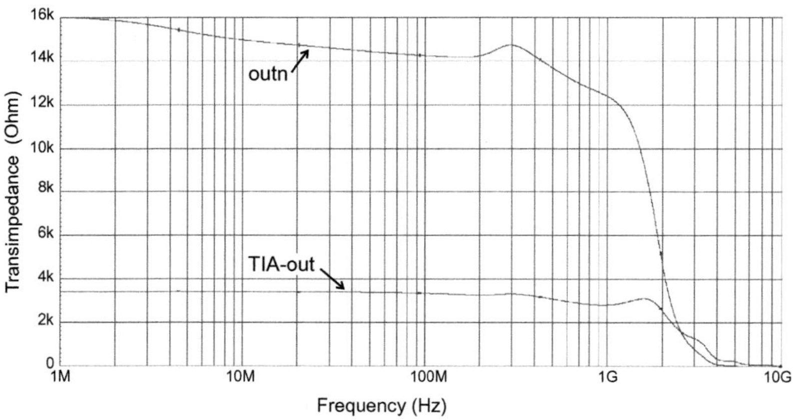

Fig. 9.21. Simulated ac analysis of three-stage TIA

any more, due to short gate lengths and high doping of the transistors. The third stage was carried out equally to the second one. P5 with a width of 13 µm and P6 which is five times as wide, are used as the variable feedback. The drain–source resistances of these transistors are adjustable by their gate voltage and therefore the transimpedance is continuously variable to achieve a high input-current range. P7/P8 build a source follower to take care of the differential amplifier load.

The simulated ac frequency response for the output of the whole chip and the TIA itself is shown in Fig. 9.21. A bandwidth of 1.32 GHz at the chip output node is achieved. The TIA itself shows a bandwidth of 2.05 GHz.

For a low optical input power and a high data rate, the open-loop gain of the amplifier must be maximum at maximum transimpedance and therefore all three amplifier stages in the TIA are active. For high optical input power and therefore large input currents the transimpedance has to be lower to prevent overdrive and therefore the feedback transistor P6 is switched on and the feedback resistance is reduced to 102 Ω. To keep the bandwidth nearly constant and avoid oscillation with low transimpedance, the open-loop gain $A_0 = 37.16\,\text{dB}$ of the amplifier in (9.1) has to be reduced

$$f_{3\,\text{dB}} = \frac{A_0 + 1}{2\pi R_F C_T}. \tag{9.1}$$

Therefore the stages two and three (P3, N3 and P4, N4) are short circuited by the switching transistor S1 and A_0 is reduced to 11.5 dB. To prevent the influence of the deactivated stages, they are switched off by setting the gate voltages of N3 and N4 to zero. For fine-tuning the amplifier's gain, the gate voltage of N2 can be modified. Figure 9.22 shows the layout plot of the TIA chip.

The extinction-ratio of the lasers used for PON is defined as 10. The laser used for the measurement, however, had an extinction ratio of 3. Therefore a correction factor of −2.1 dB for the measured sensitivity follows for the demanded specification (see Sect. 7.1.3). The sensitivities are determined at BERs of 10^{-9}, 10^{-10}, and 10^{-12} for better comparison with results in the literature.

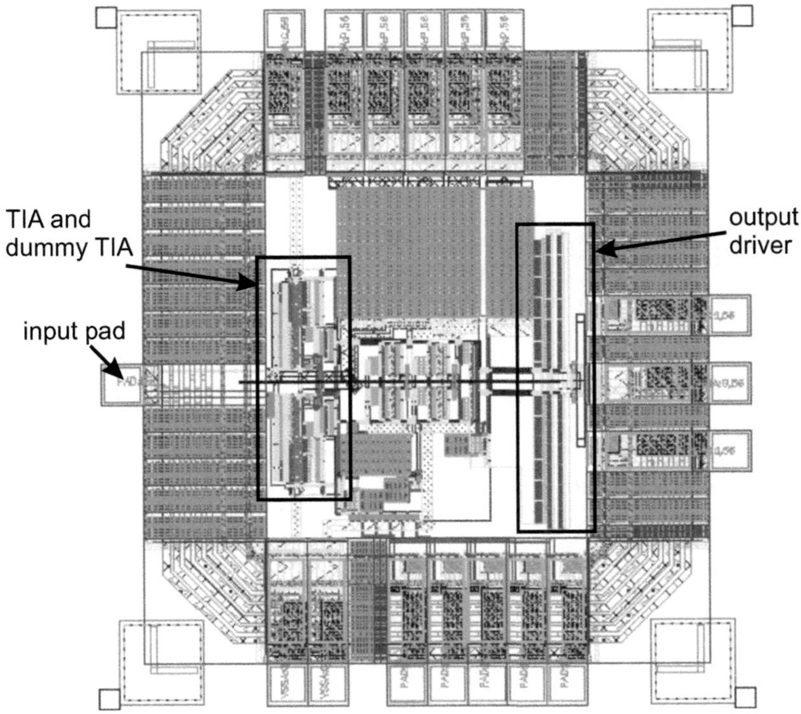

Fig. 9.22. Layout plot of three-stage TIA

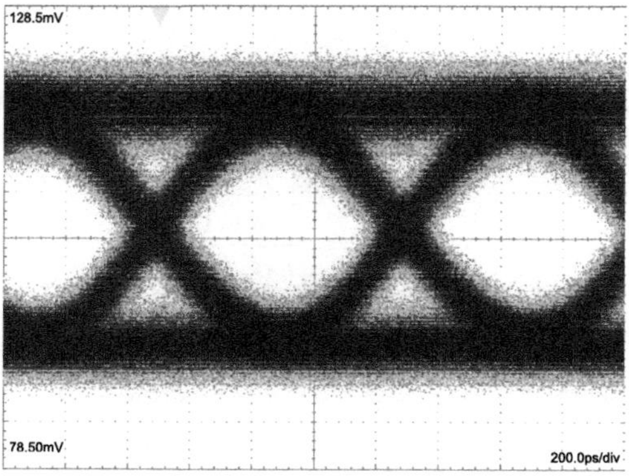

Fig. 9.23. Eye diagram of three-stage TIA for 1.25 Gb s^{-1} at an average optical input power of −25.1 dBm with EX = 3 and therefore a corrected sensitivity of −27.2 dBm (BER = 10^{-10}, EX = 10, λ = 1,310 nm, PRBS = $2^{31} - 1$)

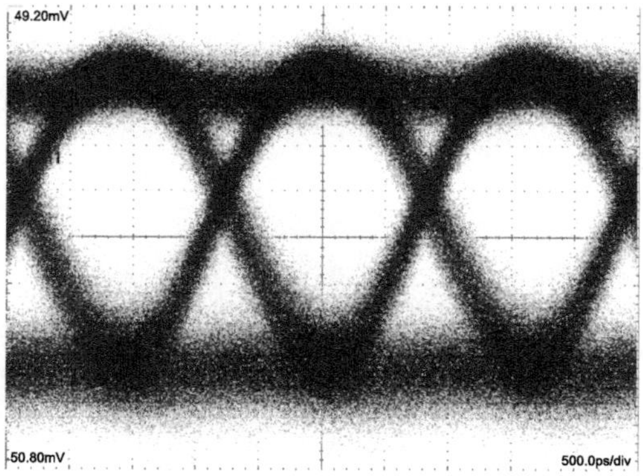

Fig. 9.24. Eye diagram of three-stage TIA for 622 Mb s^{-1} at an average optical input power of −29.2 dBm for EX = 3 and therefore a corrected sensitivity of −31.4 dBm (BER = 10^{-10}, EX = 10, λ = 1,310 nm, PRBS = $2^{31} - 1$)

In Fig. 9.23, the eye diagram for 1.25 Gb s^{-1} and an average optical input power of −25.2 dBm is shown. This value and the −29.3 dBm at 622 Mb s^{-1} (see Fig. 9.24) for the average optical input power lead to the corrected sensitivities of −31.4 dBm at a data rate of 622 Mb s^{-1} and −27.3 dBm at 1.25 Gb s^{-1} and a BER of 10^{-10} with a PRBS of $2^{31}-1$. For a BER of 10^{-9} and a PRBS of $2^{31} - 1$ the sensitivities increase to −32.4 dBm at 622 Mb s^{-1} and −28.8 dBm at 1.25 Gb s^{-1}. For a BER of 10^{-12} the sensitivities are −29.6 dBm at 622 Mb s^{-1} and −25.7 dBm at 1.25 Gb s^{-1}.

The active area of the TIA, dummy TIA and the differential stage shown in Fig. 9.8 consumes 0.0578 mm^2. The whole chip area is 1.05 mm^2. The big difference between these two values results from on-chip blocking capacitors and the 50 Ω output driver circuit. The power consumption of the circuit depends on the gain setting of the TIA. For the highest sensitivity, where all TIA stages are active, the whole chip consumes 87 mW with a supply voltage of 1.5 V. The actual receiver consumes about 36% of this amount. The rest is dissipated in the driver, which is not necessary in an SoC. For larger photocurrents and lower sensitivity the overall power consumption decreases to 73.5 mW, because the second and third amplifier stage of the TIA are switched off. The power consumption of the first amplifier stage is about 15.5 mW. The eye diagram for an average optical power of −4.5 dBm at 1.25 Gb s^{-1} and PRBS of $2^{31}-1$ is wide open and the Q-factor exceeds 10 which means a BER considerably better than 10^{-15}. This optical power was the maximum power the used laser could provide. Table 9.4 shows an overview of the measured results.

9.3.4 Three-Stage Burst-Mode Transimpedance Amplifier with Internal Feedback

The details of the circuit, designed in 120 nm CMOS technology, are shown in Fig. 9.25. The transistors N1 and P1, both with a large width-to-length ratio, build the first amplifier stage. The inverter structure was chosen to minimize the equivalent input-noise current of the amplifier. P2/N2 and P3/N3 build the second and third amplifier stage, respectively. The feedback structure consists of two transistors P4 and P5 which form the transimpedance and an additional load for the amplifier consisting of P6, N4, and N5. This feedback structure (P4′ to P6′, N4′ and N5′) is also used across the second amplifier stage. The node voltages of the input and output of the overall feedback and the node voltages of the feedback of the second stage are essentially the same and therefore the behavior of the feedbacks is the same [145]. The second and third stages use the load transistors N2 and N3, respectively. Compared to typical CMOS inverters, this allows varying the load by varying the gate voltage of N2 and N3. The additional noise of this structure compared to the typical CMOS inverter is not critical, because it is not directly at the input node and therefore is attenuated by the gain of the previous stage. P7/P8 build a source-follower to take care of the loading by the differential amplifier [146].

Table 9.4. Summarized results of three-stage TIA (EX = 10, $\lambda = 1310$ nm, PRBS $= 2^{31}-1$)

bit rate	sensitivity (dBm)	BER
1.25 Gb s^{-1}	−25.7	10^{-12}
622 Mb s^{-1}	−29.6	10^{-12}
1.25 Gb s^{-1}	−27.3	10^{-10}
622 Mb s^{-1}	−31.4	10^{-10}
1.25 Gb s^{-1}	−28.8	10^{-9}
622 Mb s^{-1}	−32.4	10^{-9}

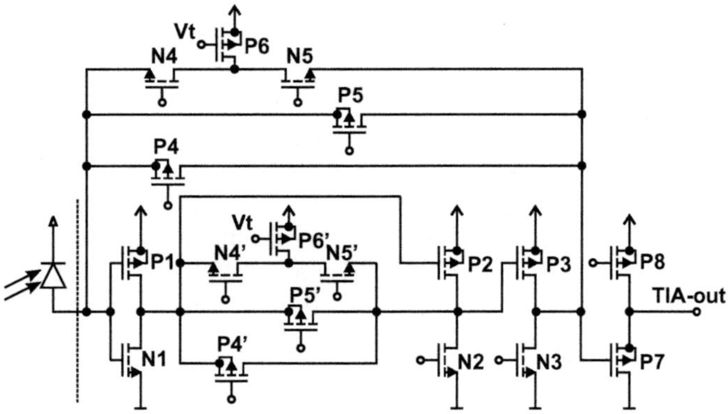

Fig. 9.25. Schematic of three-stage BM TIA with internal feedback

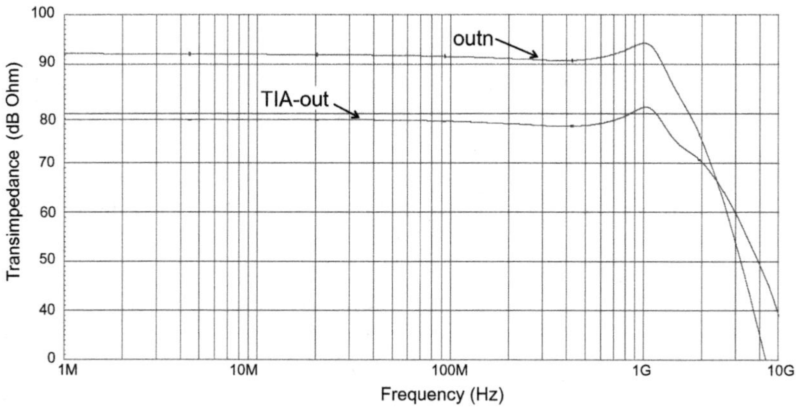

Fig. 9.26. Simulated ac analysis of three-stage BM TIA with internal feedback

The simulated ac frequency response for the output of the whole chip and the TIA itself is shown in Fig. 9.26. A bandwidth of 1.25 GHz at the chip output node is achieved. The TIA itself shows a bandwidth of 1.37 GHz.

For a low optical input power and a high data rate, the gain of the amplifier must be maximum at maximum transimpedance; only P4 and P4′ are activated. For high optical power and therefore large input currents, the transimpedance has to be lower to prevent overdrive and pulse width distortion and therefore the feedback transistor P5 is turned on. The resistance of the transimpedance can be continuously varied with the gate voltage of the transistors P4 and P5, P4′ and P5′, respectively. With P5 also P5′ is turned on and the gain A2 of the second amplifier stage is decreased. For highest optical input power the transistors N4, N4′, N5, N5′, P6, and P6′ are activated. They build a connection to VDD via P6 with a tunable gate-voltage V_t and form an extra load at the output of the amplifier and therefore decrease the gain too. N4/N5 and P6 form the load for the third stage and N4′/N5′ and P6′ for the first and second stage of the amplifier. It is possible to vary the load via V_t. The open-loop gain of the TIA can be varied from 794 for the highest transimpedance with highest sensitivity to 6.9 for the maximum optical input power and therefore the lowest transimpedance to avoid oscillation for low transimpedance. Figure 9.27 shows the layout plot of the TIA chip.

Sensitivities of −31.3 dBm at a data rate of 622 Mb s^{-1} and −28.6 dBm at 1.25 Gb s^{-1} (see eye diagrams in Figs. 9.28 and 9.29, respectively) and a BER of 10^{-10} with a PRBS of $2^{31} - 1$ are achieved. For a BER of 10^{-9} and a PRBS of $2^{31} - 1$ the sensitivity increases to −31.8 dBm at 622 Mb s^{-1} and −29.1 dBm at 1.25 Gb s^{-1}. The active die area of the TIA, dummy TIA and the differential stage consumes 0.032 mm^2. The whole chip area is 1.05 mm^2. The whole chip consumes 88.5 mW with a power supply voltage of 1.5 V. The eye diagram for an optical power of −1.6 dBm at 1.25 Gb s^{-1} and PRBS of $2^{31} - 1$ is wide open and the BER is considerably better than 10^{-15}.

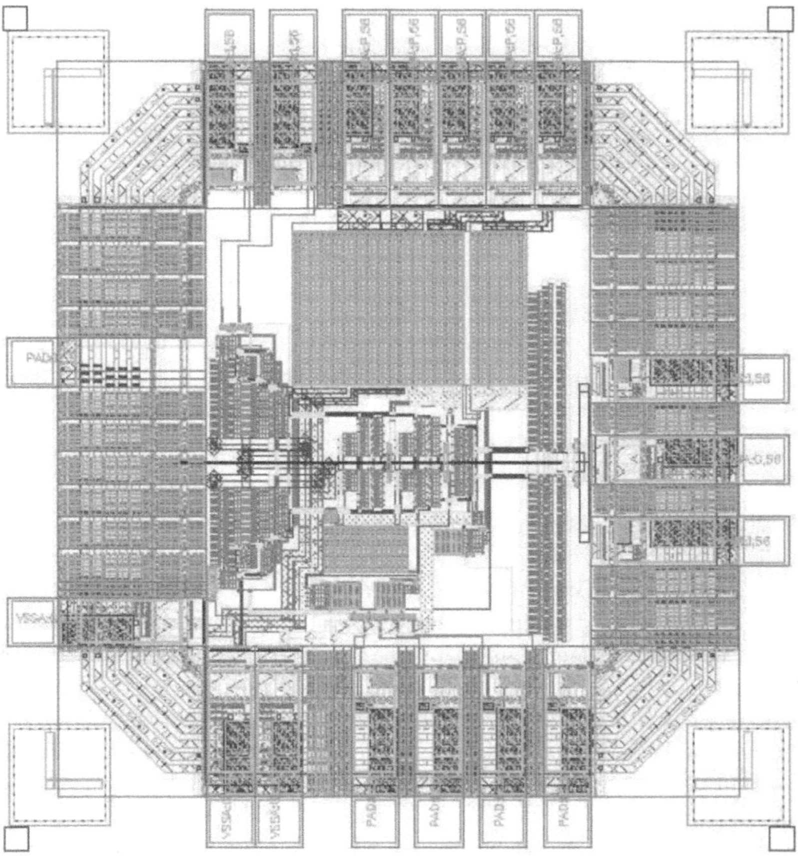

Fig. 9.27. Layout plot of three-stage BM TIA with internal feedback

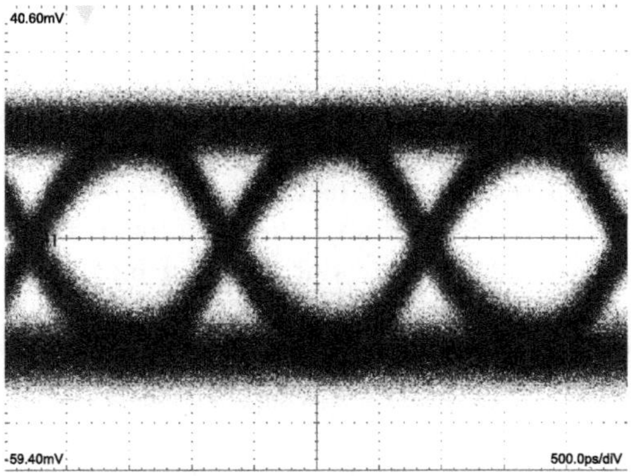

Fig. 9.28. Eye diagram of three-stage BM TIA with internal feedback at 622 $\mathrm{Mb\,s^{-1}}$ for an average optical input power of -31.3 dBm (BER $= 10^{-10}$, EX $= 10$, $\lambda = 1310$ nm, PRBS $= 2^{31} - 1$)

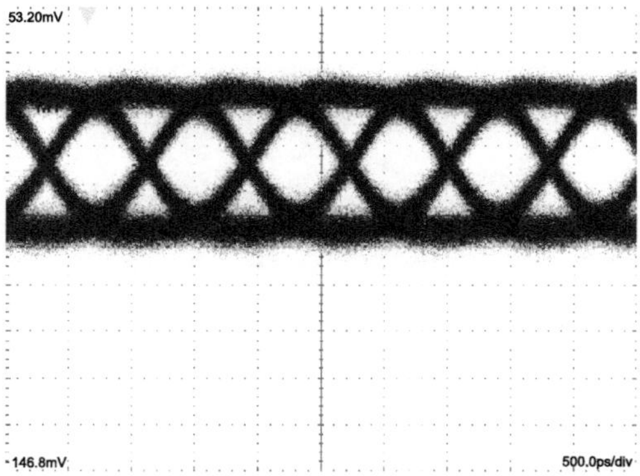

Fig. 9.29. Eye diagram of three-stage BM TIA with internal feedback at 1.25 $\mathrm{Gb\,s^{-1}}$ for an average optical input power of -28.6 dBm (BER $= 10^{-10}$, EX $= 10$, $\lambda = 1310$ nm, PRBS $= 2^{31} - 1$)

This design also works at higher frequencies due to the tuneable feedback. But the sensitivity is quite poor with -14.5 dBm at 2.5 $\mathrm{Gb\,s^{-1}}$ with a BER of 10^{-10} and -15.5 dBm with a BER of 10^{-9}, respectively.

Figure 9.32 shows the simulated result at the TIA output node for switching from the highest transimpedance to the lowest transimpedance and also switching the input signal from minimum photocurrent swing of 2 μA_{PP} to the maximum photocurrent swing of 300 μA_{PP}. The dc level moves about 100 mV when the gain is switched. The switching and settling is finished within 13.4 ns for switching from maximal transimpedance to minimal transimpedance and in 12.4 ns the other way round [147]. The switching is simulated for a "0"–"1"–"0" bit sequence it can be seen in Fig. 9.32. For proper measurement of the switching time the layout of this chip needs a redesign for faster providing

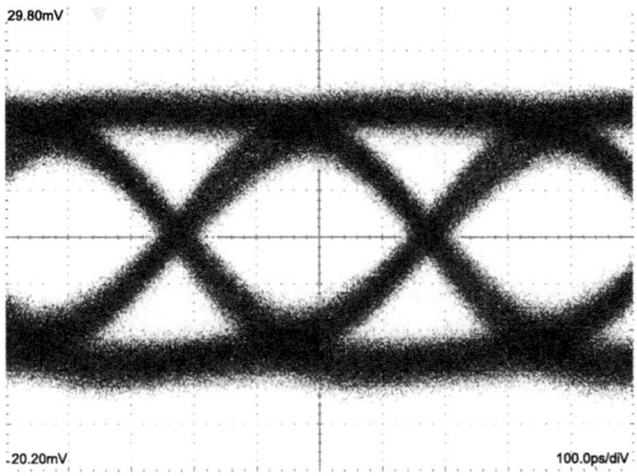

Fig. 9.30. Eye diagram of three-stage BM TIA with internal feedback at $2.5\,\mathrm{Gb\,s^{-1}}$ for an average optical input power of $-14.5\,\mathrm{dBm}$ (BER $= 10^{-10}$, EX $= 10$, $\lambda = 1{,}310\,\mathrm{nm}$, PRBS $= 2^{31} - 1$)

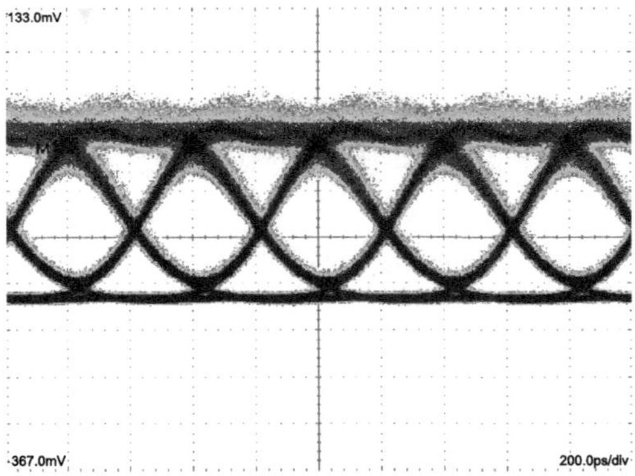

Fig. 9.31. Eye diagram of three-stage BM TIA with internal feedback at $2.5\,\mathrm{Gb\,s^{-1}}$ for an average optical input power of $-1.6\,\mathrm{dBm}$

the bias voltage. There is no measurement of the switching due to the fact that no power switching laser is available jet.

Table 9.5 gives an overview of the measured results of the three-stage BM TIA with internal feedback.

9.3.5 Three-Stage Burst-Mode Transimpedance Amplifier for $2.5\,\mathrm{Gb\,s^{-1}}$

The circuit of this TIA is similar to the design of Sect. 9.3.3. It is again designed in 120 nm CMOS technology for a higher bit rate of $2.5\,\mathrm{Gb\,s^{-1}}$ and for fast switching between high and low transimpedance. The main difference to the structure of the other designs is that there is a reference voltage V_{ref} instead of the dummy TIA at the negative input of the differential amplifier (see Fig. 9.33) [94, 148]. The reason for this design with reference voltage is

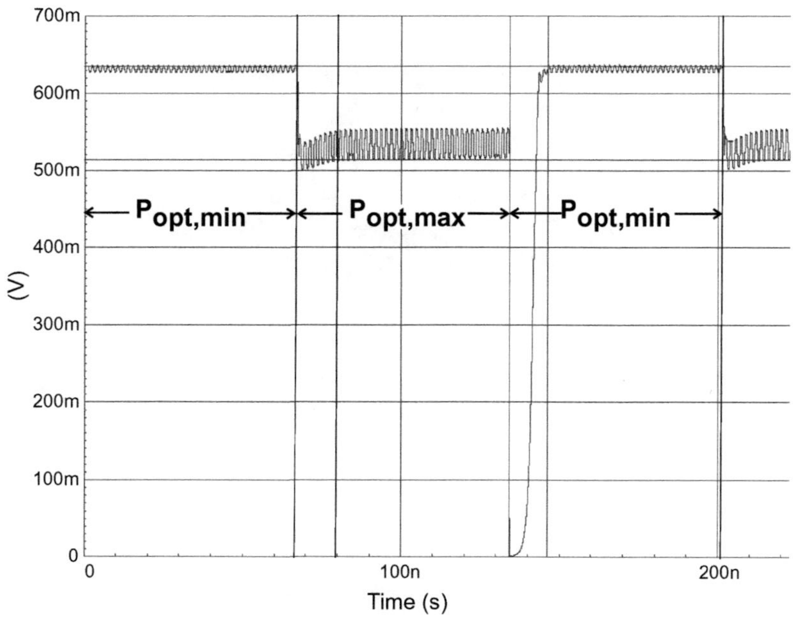

Fig. 9.32. Simulated switching of three-stage BM TIA with internal feedback from minimal to maximal optical input power at 1.25 Gb s^{-1}

Table 9.5. Summarized results of three-stage BM TIA with internal feedback (EX = 10, $\lambda = 1310$ nm, PRBS = $2^{31} - 1$)

bit rate	sensitivity (dBm)	BER
1.25 Gb s^{-1}	−27.3	10^{-12}
622 Mb s^{-1}	−30.0	10^{-12}
1.25 Gb s^{-1}	−28.6	10^{-10}
622 Mb s^{-1}	−31.3	10^{-10}
1.25 Gb s^{-1}	−29.1	10^{-9}
622 Mb s^{-1}	−31.8	10^{-9}
2.5 Gb s^{-1}	−14.5	10^{-10}
2.5 Gb s^{-1}	−15.5	10^{-9}

that the switching of the open-loop gain, also leads to a change of the dc operating point at the TIA output. The dummy TIA has a much lower bandwidth to suppress the additional noise of the differential design and therefore the settling time of variations in the dc-operating point is long. This can be avoided, if a fast switching reference voltage is fed to the second input of the differential amplifier.

Another advantage of this design is that the power consumption is reduced by the consumption of the dummy TIA, compared to the previous designs.

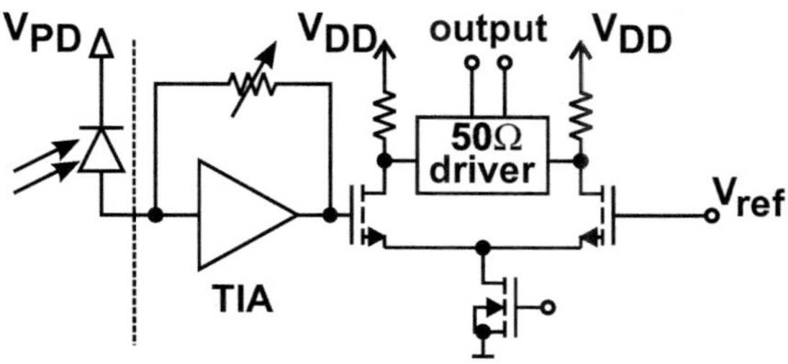

Fig. 9.33. Basic schematic of the TIA chip for 2.5 Gb s^{-1}

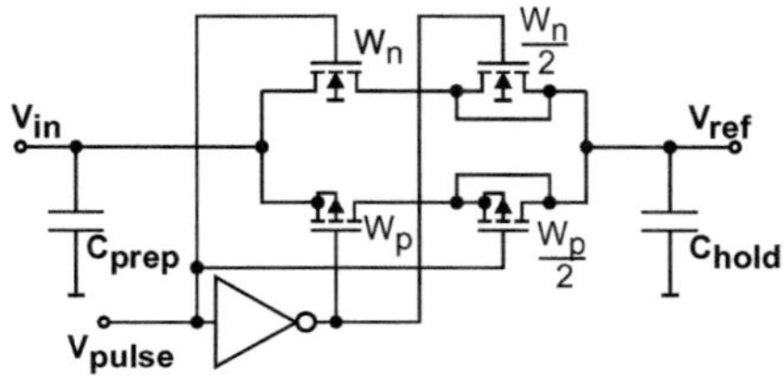

Fig. 9.34. Prepare and hold circuit for fast switching

This is quite a lot, because the TIA consumes about 100 mW to achieve the high bandwidth and to show a high sensitivity.

In the final optical burst-mode receiver V_{ref} is provided by a digital-to-analog converter (DAC). For very fast gain switching of the TIA it is necessary to provide the bias and reference voltages even faster (<1 ns). Due to the fact that the external biasing, as well as the internal DAC, do not meet the requirements in switching speed the prepare and hold circuit, shown in Fig. 9.34, is inserted. This circuit separates the external voltage from the voltage fed into the circuit and therefore enables slow change of the external voltage and fast switching for the chip.

The voltage from the DAC equals V_{in} in the circuit. The biasing and also the reference voltage are provided off-chip for the test chip, due to the fact that there is no digital signal processing part on the test chip. V_{in} is prepared at the prepare capacitance C_{prep}. The switch between the capacitances C_{prep} and C_{hold}, consisting of the two n-MOS- and two p-MOSFETs, is open during the burst circle. During this time V_{ref} is constant and at V_{in} the next voltage is prepared. When the prepared voltage is switched to V_{ref} the MOSFETs connect through and V_{ref} approximately equals V_{in}. The actual charge of C_{hold} has to be taken into consideration for the choice of V_{ref} to achieve the desired V_{in}. The switch opens again and V_{in} can be prepared to a different voltage. The voltage at V_{ref} is "stored" at the hold capacitance C_{hold}. The input V_{Pulse} is generated on chip with an monostable circuit shown in the middle of Fig. 9.35 [149]. With the rising edge at the input voltage *Switch* a pulse is generated, which copies V_{in} to V_{ref} in Fig. 9.34. Every bias voltage of the TIA chip is provided the same way (see Fig. 9.35).

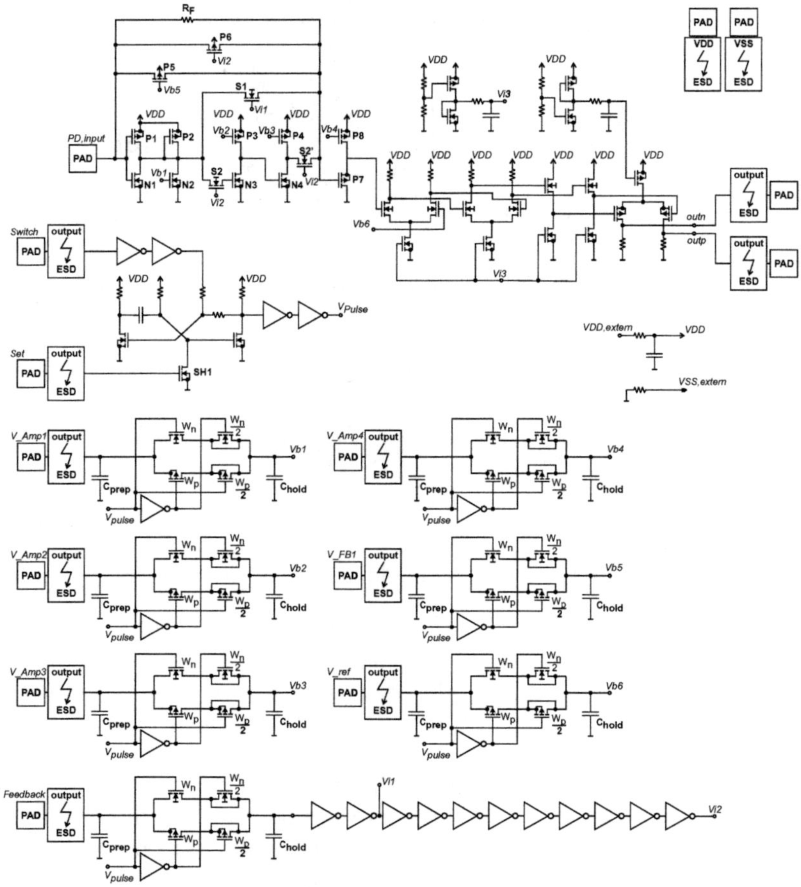

Fig. 9.35. Complete circuit of three-stage BM TIA for 2.5 Gb s^{-1}

This circuit has two advantages: it enables switching faster than the settling time of the DAC or an external voltage, respectively, and the noise of the circuits behind the open switch is cut off and therefore the noise performance is good.

To enable a static operation, the selection pin *Set* is inserted. It switches the transistor SH1 (see Fig. 9.35) on and therefore the voltage V_{Pulse} is set to VDD, the switch in the prepare-and-hold circuit is closed and V_{in} is connected to V_{ref} (see Fig. 9.34). For more details concerning the prepare-and-hold circuit see [94].

The switch is designed with n-MOS and p-MOSFETs in parallel to enable switching of every voltage between VDD and VSS. The switching transistors are followed by short circuited transistors with the half gate width and the inverse input signal to minimize load injection of the switches.

Again the TIA consists of a three-stage amplifier. For the lowest optical input power, all three amplifier stages are active to achieve the gain necessary to reach the bandwidth despite the high feedback resistance and the high input node capacitance. The feedback consists of an actual polysilicon resistor ($R_F = 7\,\mathrm{k\Omega}$) and two parallel p-MOS transistors (P5 and P6) with adjustable gate voltage to vary the feedback resistance down to $100\,\Omega$ if necessary. P1 and N1 form the first amplifier stage, P2 forms a constant diode load for the first stage, and N2 forms a variable load to increase the bandwidth of the first stage and to improve stability.

To keep the system stable when the transimpedance is lowered it is again necessary to lower the open-loop gain of the amplifier. Therefore the second (P3/N3) and the third (P4/N4) amplifier stage are bypassed by the switch S1 (see Fig. 9.36). The transistors S2 and S2′ separate the bypassed stages from the network to minimize their influence.

The whole chip consumes an area of $1 \times 1.5\,\mathrm{mm}^2$. The area of the actual TIA is only $0.019\,\mathrm{mm}^2$. The difference is consumed by the pads, the output driver, and filtering of the supply voltage. Figure 9.38 shows the layout plot of the chip. The power consumption of the chip is 293 mW, where the TIA itself consumes 99 mW. The chip is again mounted on a printed circuit board for the characterization. Eye-catching is the input pad of the TIA chip, which is on the left-hand side, the pad which is moved towards the core of the circuit. The reason therefore is the fact that the closed VDD and VSS rings in the pad ring offer only a metal 2 and metal 3 connection to the bond-pad outside these structures and therefore the RC constant of this very long connection of more than 100 µm would be disadvantageous. Therefore, the pad itself is moved next to the core with a very short connection of under 1 µm. The closed pad ring is still faultless, because the wiring is closed outside the input pad.

AC analysis of the whole chip delivers a bandwidth of 2.11 GHz. The measured sensitivities are $-27.59\,\mathrm{dBm}$ at $1.25\,\mathrm{Gb\,s^{-1}}$ and a BER of 10^{-10}, $-22.55\,\mathrm{dBm}$ at $2\,\mathrm{Gb\,s^{-1}}$ and $-20.41\,\mathrm{dBm}$ at $2.5\,\mathrm{Gb\,s^{-1}}$ respectively (see

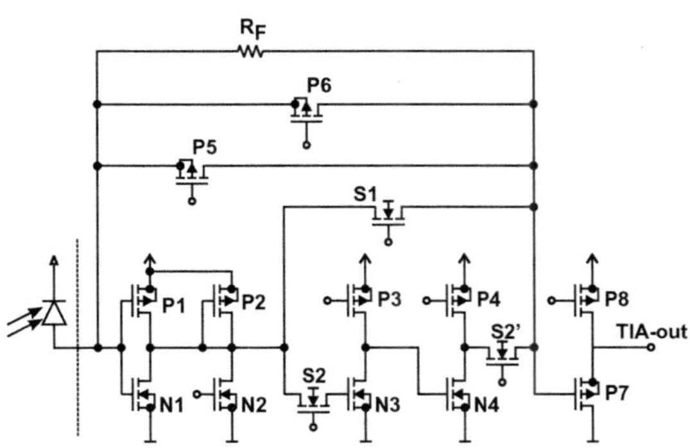

Fig. 9.36. Schematic of three-stage BM TIA for $2.5\,\mathrm{Gb\,s^{-1}}$

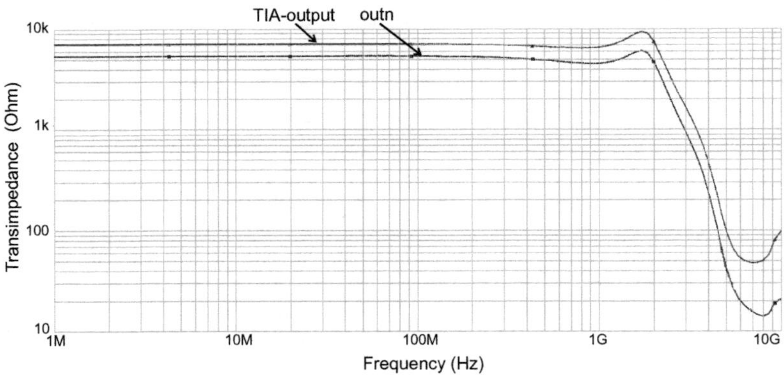

Fig. 9.37. Frequency response of three-stage BM TIA for 2.5 Gb s^{-1}

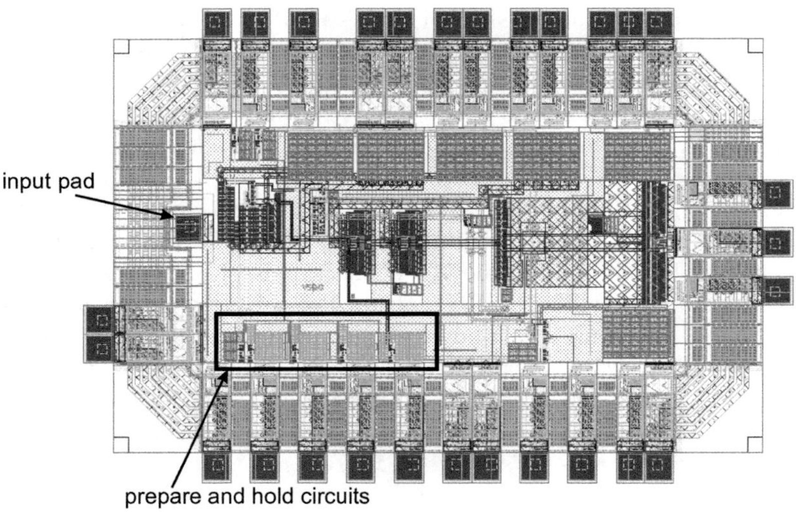

Fig. 9.38. Layout plot of three-stage BM TIA for 2.5 Gb s^{-1}

Figs. 9.39 to 9.41). An overview of the results for a BER of 10^{-9}, 10^{-10}, and 10^{-12} is given in Table 9.6.

The most important feature of this optical receiver is the fast gain switching. The receiver shows a switching time of 0.8 ns for switching from maximal transimpedance to minimal transimpedance, and 2.3 ns the other way round. Figure 9.42 shows the simulated result. It shows the voltage at the TIA output. The settling of the TIA is finished in maximal six bits at the output of the whole chip, for a data rate of 2.5 Gb s^{-1}. Again, a simulated result is presented, because the power switching laser has to be built next.

Due to the fact that this design uses the metal layers up to the top metal layer, the micrograph of the test chip shows except for the pads only some

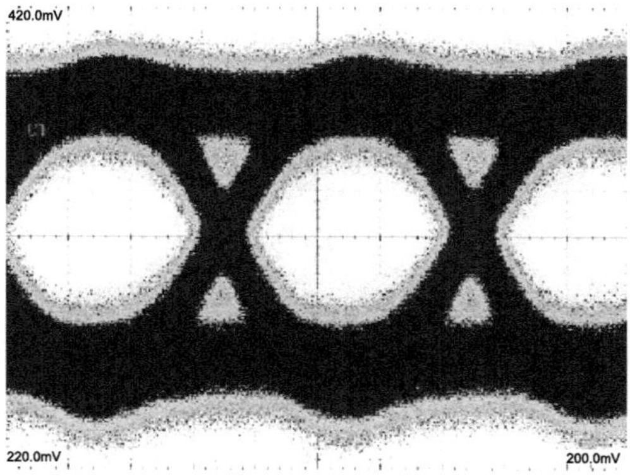

Fig. 9.39. Eye diagram of three-stage BM TIA for 2.5 Gb s^{-1} for 1.25 Gb s^{-1} at an average optical input power of −27.6 dBm (BER = 10^{-10}, EX = 10, λ = 1310 nm, PRBS = $2^{31} - 1$)

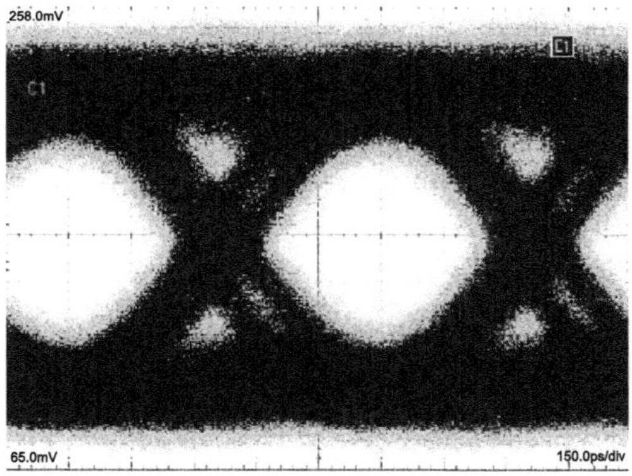

Fig. 9.40. Eye diagram of three-stage BM TIA for 2.5 Gb s^{-1} for 2 Gb s^{-1} at an average optical input power of −22.55 dBm (BER = 10^{-10}, EX = 10, λ = 1310 nm, PRBS = $2^{31} - 1$)

metal lines and therefore it is presented here. Figure 9.43 shows the microphotograph of the TIA chip. Again on the left-hand side, the input pad is visible, and on the right-hand side, the differential output pads with a VSS pad in between can be seen. On top are the supply voltages and on bottom the bias-voltage pads.

9.3.6 Three-Stage Burst-Mode Transimpedance Amplifier for Ultra-Fast Gain Switching

This optical receiver again is a differential design. No variation in the dc-operating point of the circuit was the main focus of the design, because the switching time again should be minimized. Again the bandwidth of the dummy TIA is lower than the one of the TIA itself, to keep the sensitivity high. By

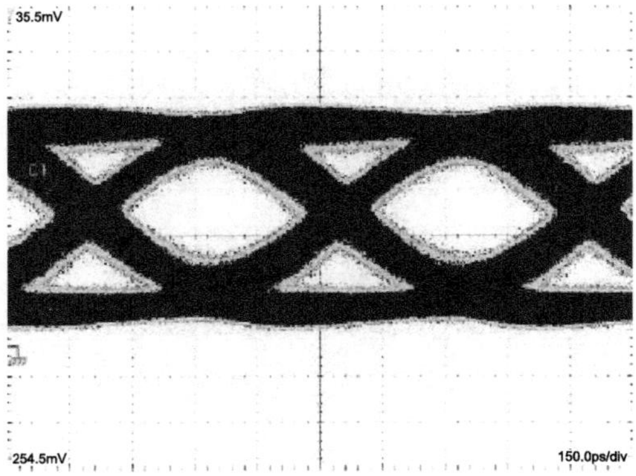

Fig. 9.41. Eye diagram three-stage BM TIA for 2.5 $\mathrm{Gb\,s^{-1}}$ at an average optical power of −20.4 dBm (BER = 10^{-10}, EX = 10, $\lambda = 1310$ nm, PRBS = $2^{31} - 1$)

Table 9.6. Summarized results of three-stage BM TIA for 2.5 $\mathrm{Gb\,s^{-1}}$ (EX = 10, $\lambda = 1310$ nm, PRBS = $2^{31} - 1$)

bit rate ($\mathrm{Gb\,s^{-1}}$)	sensitivity (dBm)	BER
1.25	−26.65	10^{-12}
2	−20.85	10^{-12}
2.5	−19.68	10^{-12}
1.25	−27.59	10^{-10}
2	−22.55	10^{-10}
2.5	−20.41	10^{-10}
1.25	−28.20	10^{-9}
2	−23.93	10^{-9}
2.5	−20.85	10^{-9}

avoiding changes in the dc-operating point, the lower bandwidth of the dummy TIA does not influence the switching performance.

The TIA again consists of a three-stage amplifier. This time three classical inverter stages are used (P1/N1, P2/N2, P3/N3), where the first and the third stage are equipped with diode loads (N1′ and N3′). For the lowest optical input power, the switches S1, S1′, S2, and S2′ are open. Again the first inverter stage is optimized regarding to noise. For the minimum optical input power the transistor P6 works as resistor and the other feedback transistors (P7 and P7′) are switched off. The circuit of the TIA is shown in Fig. 9.44.

The achieved bandwidth of the whole chip is 2.25 GHz. The TIA alone shows a bandwidth of 2.5 GHz. Figure 9.45 shows the frequency response of both signals, the output node of the chip *outn* and the TIA output.

The transient response with minimum input current is shown in Fig. 9.46. The curves show the signal at the output of the TIA (*TIA-out*), after the differential amplifier (*diff-out*) and at the output node of the whole chip (*outn*).

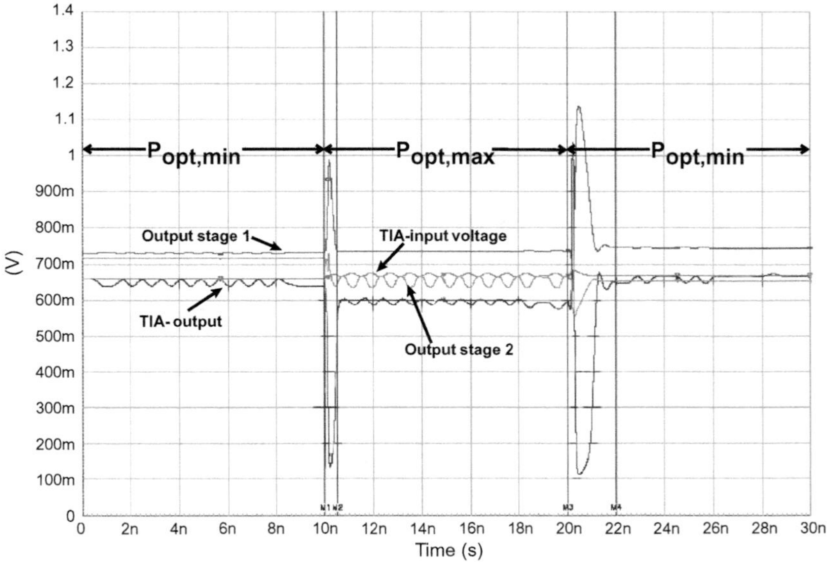

Fig. 9.42. Simulated switching of three-stage BM TIA for 2.5 Gb s^{-1} from minimal to maximal optical input power at 2.5 Gb s^{-1}

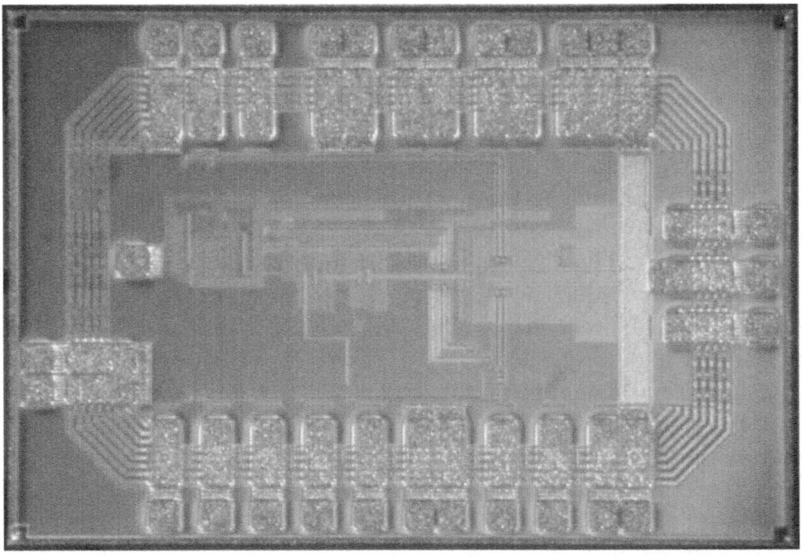

Fig. 9.43. Microphotograph of three-stage BM TIA for 2.5 Gb s^{-1}

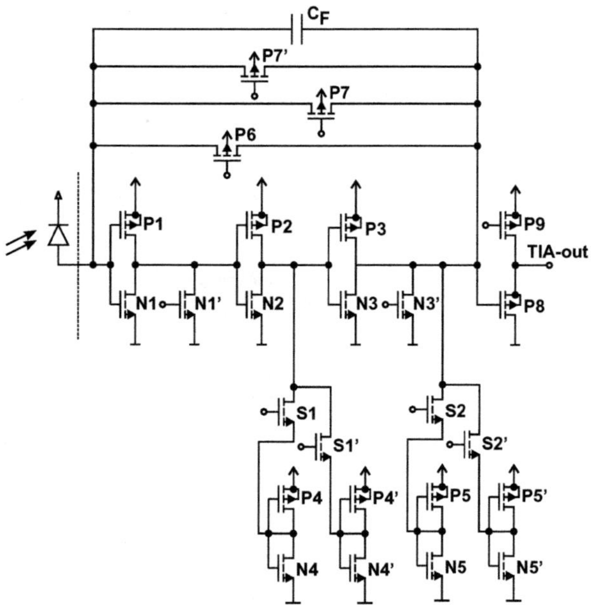

Fig. 9.44. Schematic of three-stage BM TIA for ultra-fast switching

The feedback resistance in this case is 7.97 kΩ with an open-loop gain of 61.7 dB.

All transient analyses are stimulated with a pseudorandom bit sequence at a bit rate of $2.5\,\mathrm{Gb\,s^{-1}}$. The input current of $i_{\mathrm{in}} = 1\,\mu\mathrm{A}$ is too small, due to the noise of the circuit, therefore the output voltage swing will be larger than shown in Fig. 9.46.

The noise simulation lead to an input referred noise current of 0.3 μA, which leads to an average optical input power of −26 dBm at $2.5\,\mathrm{Gb\,s^{-1}}$. With the given responsivity of $0.85\,\mathrm{A\,W^{-1}}$ the average input current swing has to be 3.88 μA. The output noise density is shown in Fig. 9.47. It shows the $1/f$ noise at low frequencies and the increase in dependance on the squared frequency at high frequencies. The drop at the end is because the gain is decreasing at high frequencies.

For higher optical input power the open-loop gain of the circuit is lowered by switching additional diode loads of the second and third stage active. This leads to a minimum open-loop gain of 25.4 dB. Switching all feedback transistors on gives the minimum feedback resistance of 185 Ω. The additional diode loads consist of an NMOS and a PMOS transistor each, which are connected together and designed that way that the potential of the output node of the inverter stage is the same as the potential of the drain node of the diode loads. This enables switching with short settling time when the open-loop gain has to be lowered.

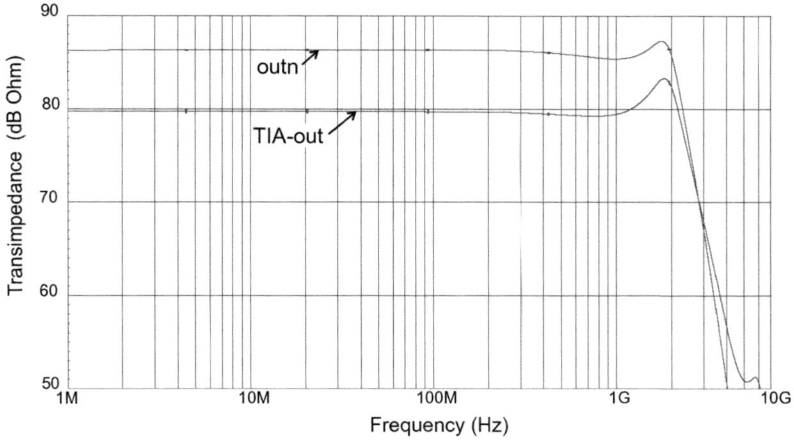

Fig. 9.45. Frequency response of three-stage BM TIA for ultra-fast switching

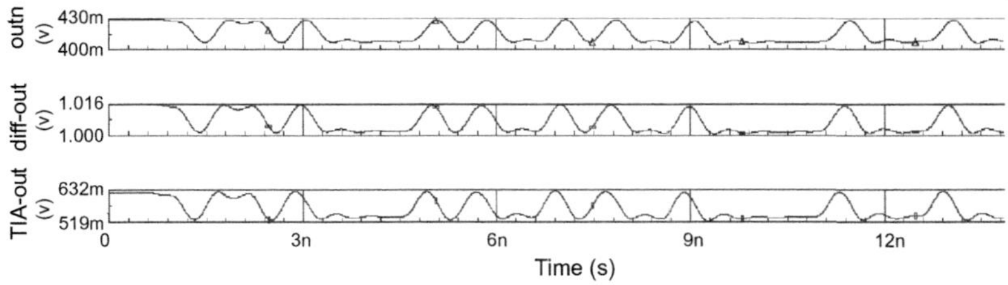

Fig. 9.46. Transient analysis for $i_{\text{in}} = 1\,\mu\text{A}$ of three-stage BM TIA for ultra-fast switching

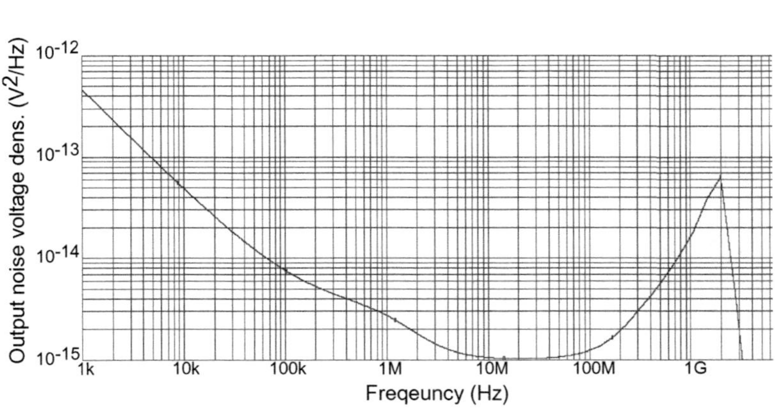

Fig. 9.47. Output noise voltage density of output node of the test chip

This design is built for three different transimpedance settings. For the lowest optical input power, as described before, only P6 is active and the switches are open. In the medium transimpedance setting, feedback transistor P7 is switched on additionally and therefore lowers the transimpedance. The open-loop gain is lowered by closing the switches S1 and S2, which activate the loads P4/N4 for the second inverter stage and P5/N5 for the third stage.

For highest optical input power and lowest transimpedance, all feedback transistors (P6, P7, and P7′) are switched on and also the switches S1, S1′, S2, and S2′ are closed and therefore all additional loads are active, too. The transistors for the switches have the same size, they have a width-to-length ratio of 254 and are not at minimum length to ensure a proper disconnection.

Figure 9.48 shows the transient response for an input current of $i_{\text{in}} = 300\,\mu\text{A}$ at $2.5\,\text{Gb}\,\text{s}^{-1}$ for the three nodes mentioned before (*TIA-out*, *diff-out* and *outn*).

The disadvantage of this circuit is that the power consumption is 227.5 mW, which is relatively high, especially compared to the design described in Sect. 9.3.4, because in this design the additional loads always consume power, even when they are not active. It is possible to use a transistor to switch off the additional loads while they are not active but it is not implemented in this design. The TIA and dummy TIA consume together 110 mW.

In the dummy TIA, all additional loads are always active and the feedback capacitance is higher, to guarantee stability. Again the noise of the dummy TIA is additionally eliminated by an RC-filter.

Due to planarization and passivation, again a layout plot is presented in Fig. 9.50. This is also the reason why simulated results are presented. The whole chip consumes an area of $1 \times 1.5\,\text{mm}^2$. The TIA, dummy TIA and differential amplifier consume $0.059\,\text{mm}^2$. This area rises to $0.0764\,\text{mm}^2$ when the prepare-and-hold circuits for the bias voltages are added. The TIA also works without the prepare-and-hold circuits due to the fact that the

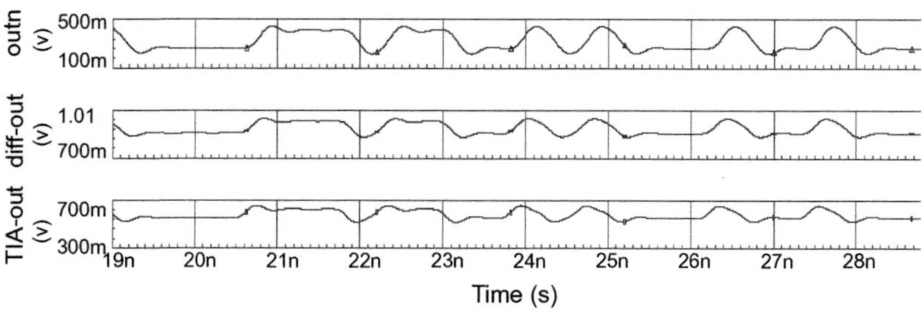

Fig. 9.48. Transient analysis for $i_{\text{in}} = 300\,\mu\text{A}$ of three-stage BM TIA for ultra-fast switching

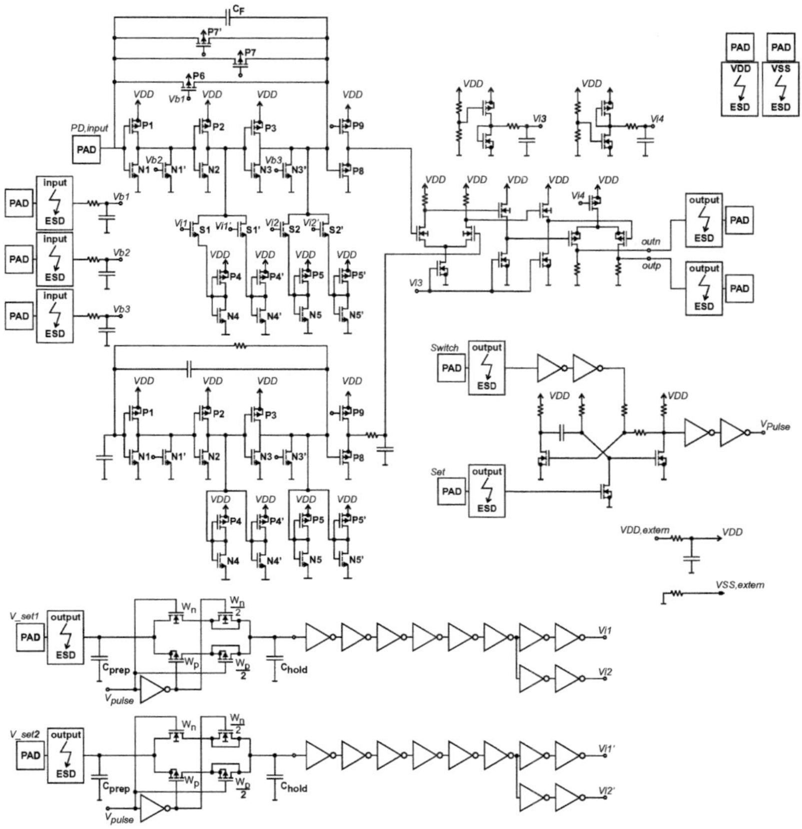

Fig. 9.49. Complete circuit of three-stage BM TIA for ultra-fast switching

incoming bias voltages are digital voltages and therefore fed into inverter lines for edge shaping but the prepare-and-hold circuits are implemented to get a well-defined switching time.

The switches in this design are all digital. There are three analog bias voltages, which are not switched. To achieve a sharp edge for switching, the incoming signal is routed via several inverting stages, which also offer the opportunity for delays between the actual switching of the feedback and the additional loads.

The switching of this design is finished in 0.83 ns from minimal optical input power to maximal optical input power and 1.64 ns back again to the minimum optical input power. Figure 9.51 shows the transient analysis of the switching.

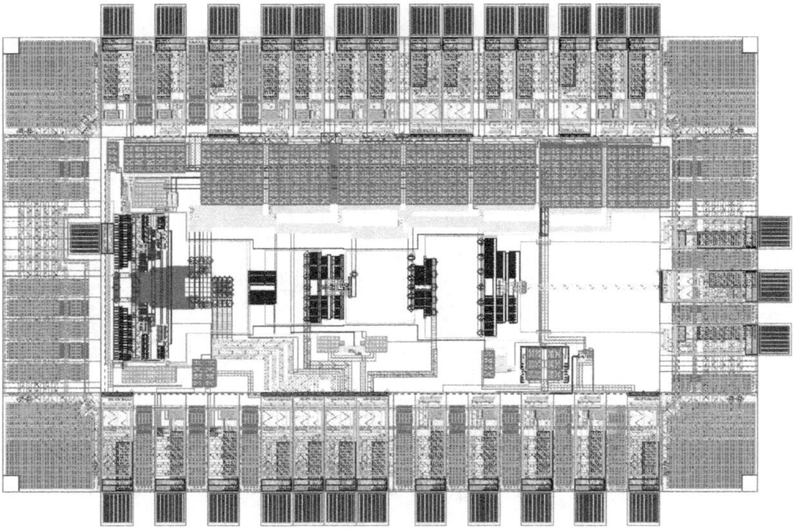

Fig. 9.50. Layout plot of three-stage BM TIA for ultra-fast switching

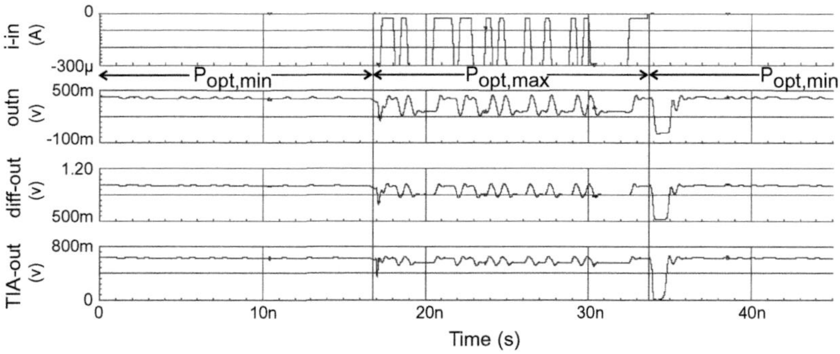

Fig. 9.51. Switching behavior of three-stage BM TIA for ultra-fast switching

9.4 Summary and Comparison

Table 9.7 summarizes the results achieved with our own designs in deep-sub-micron CMOS. The sensitivities of the designs are given for $622\,\mathrm{Mb\,s^{-1}}$, $1.25\,\mathrm{Gb\,s^{-1}}$, and $2.5\,\mathrm{Gb\,s^{-1}}$, if available. All results are given for a $\mathrm{PRBS} = 2^{31} - 1$ and a $\mathrm{BER} = 10^{-10}$. It should be mentioned that the results of the last test chip described in Sect. 9.3.6 are simulated results due to the fact that the chip was not yet available. Table 9.8 summarizes all results mentioned in this book.

Figure 9.52 shows an overview of all mentioned results up to $3\,\mathrm{Gb\,s^{-1}}$. Comparing the sensitivities, only three results at $1.25\,\mathrm{Gb\,s^{-1}}$ are better than the achieved sensitivities of our own chips at this data rate. The first one [5]

Table 9.7. Result overview of our own designs in 0.18 μm and 0.12 μm CMOS ($\lambda = 1.3\,\mu$m)

circuit section	sens. @ 622 Mb s^{-1} (dBm)	sens. @ 1.25 Gb s^{-1} (dBm)	sens. @ 2.5 Gb s^{-1} (dBm)	power cons. (mW)	in. pow. range @ 1.25Gb s^{-1}	switch. time (ns)
folded-cascode TIA Sect. 9.3.1	−30.7	−23.7		97.2		
three-stage TIA Sect. 9.3.3	−31.4	−27.3		87	22.8	
three-inverter TIA Sect. 9.3.2	−29.9	−28.2		154.5	19.3	
three-stage BM TIA w. int. feedb. Sect. 9.3.4	−31.3	−28.6	−14.5	88.5	27	13.4[a]
three-stage BM TIA for 2.5 Gb s^{-1} Sect. 9.3.5		−27.6	−20.4	293	20	2.3[a]
BM TIA for ultra-fast switch. Sect. 9.3.6			−26[a]	227.5[a]	19[a]	1.64[a]

[a] Simulated result

reported a *simulated* sensitivity of −29 dBm. Compared to our best own *measured* result of −28.6 dBm this value is only slightly better, but experience showed that the measured sensitivities always were behind the simulated ones, about 1 dBm at least, due to parasitic effects in the layout and finite power supply rejection for inverters. The other result of [6] is measured with an avalanche photodiode with a multiplication factor of 6 and therefore has an advantage of 7.8 dB over a pin photodiode. [107] achieved a sensitivity of −34 dBm at 1.5 Gb s^{-1} but also used an APD, which has a multiplication factor of 10 resulting in 10 dB advantage over the pin photodiode. The sensitivity advantage of the APD has to be paid for by an additional high and stabilized supply voltage. Considering that, also this design is comparable with our own work.

At 2.5 Gb s^{-1} the measured sensitivities of our own work, especially that of the burst-mode receiver described in Sect. 9.3.5, are around and even better than the state of the art. The TIA chip from Sect. 9.3.4 is not really highly sensitive at 2.5 Gb s^{-1}, but the design was done for 1.25 Gb s^{-1} and works at the higher data rate with a medium gain setting. The measured results

Table 9.8. Summary of sensitivities of state of the art and our own designs

reference	process	sensitivity	power cons. (mW)	in. pow. range (dB)
[105]	0.6 μm BiCMOS	−29.4 dBm @ 622 Mb s^{-1}		
[4]	Si bipolar	−27.7 dBm @ 1.25 Gb s^{-1}		21
[107]	Si bipolar	−34 dBm @ 1.5 Gb s^{-1} [a]	110	15
[108]	0.8 μm Si BiCMOS	−23.3 dBm @ 1.7 Gb s^{-1}		
[109]	Si bipolar	−12.5 dBm @ 1.25 Gb s^{-1}	865	
[109]	Si bipolar	−11 dBm @ 2 Gb s^{-1}	865	
[110]	0.5 μm Si bipolar	−19.4 dBm @ 2.5 Gb s^{-1}		17.4
[111]	Si bipolar	−24 dBm @ 1.9 Gb s^{-1}	210	
[111]	Si bipolar	−20 dBm @ 3.5 Gb s^{-1}	210	
[112]	Si bipolar	−11 dBm @ 5 Gb s^{-1}	700	
[113], [114]	0.25 μm BiCMOS	−17 dBm @ 10 Gb s^{-1}	140	
[115]	0.3 μm Si bipolar	−18.1 dBm @ 10 Gb s^{-1}		20
[116]	0.5 μm Si bipolar	−21.5 dBm @ 15 Gb s^{-1} [b]	9	
[117]	0.25 μm Si bipolar	−16.4 dBm @ 10 Gb s^{-1}	41	
[117]	0.25 μm Si bipolar	−15.5 dBm @ 11.25 Gb s^{-1}	41	
[117]	0.25 μm Si bipolar	−14.5 dBm @ 12 Gb s^{-1}	41	
[118]	0.8 μm Si bipolar	−22.5 Gb s^{-1} @ 13 Gb s^{-1}	280	
[119]	0.8 μm SiGe BiCMOS	−27 dBm @ 155 Mb s^{-1}	500	26.0
[6]	0.35 μm SiGe BiCMOS	−30.2 dBm @ 1.25 Gb s^{-1} [c]		21
[120]	0.8 μm Si-SiGe HBT	−19 dBm @ 5 Gb s^{-1}	65	
[121]	0.8 μm Si-SiGe HBT	−22 dBm @ 5 Gb s^{-1}	10	
[122]	0.35 μm SiGe bipolar	−18.4 dBm @ 10 Gb s^{-1}	117	
[123]	0.18 μm SiGe BiCMOS	−6.9 dBm @ 10 Gb s^{-1}		
[124]	SiGe HBT	−29.5 dBm @ 10 Gb s^{-1} [a]		
[127]	0.6 μm CMOS	−30.9 dBm @ 622 Mb s^{-1} [b]		
[128]	0.7 μm CMOS	−21.8 dBm @ 1 Gb s^{-1} [b]	100	
[98]	0.35 μm CMOS	−27 dBm @ 622 Mb s^{-1}	260	
[98]	0.35 μm CMOS	−22.5 dBm @ 1.25 Gb s^{-1}	260	
[129]	0.25 μm CMOS	−24 dBm @ 1.25 Gb s^{-1}	26	
[129]	0.25 μm CMOS	−14 dBm @ 1.5 Gb s^{-1}	26	
[130]	0.5 μm CMOS	−30.1 dBm @ 622 Mb s^{-1}	<30	22.2
[130]	0.5 μm CMOS	−27.4 dBm @ 1 Gb s^{-1}	<30	19.5
[130]	0.5 μm CMOS	−18.5 dBm @ 1.3 Gb s^{-1}	<30	5.5
[133]	0.25 μm CMOS	−244 dBm @ 1.25 Gb s^{-1}		21 dB

Table 9.8. Continued

reference	process	sensitivity	power cons. (mW)	in. pow. range (dB)
[130]	0.5 μm CMOS	−31.4 dBm @ 622 Mb s^{-1}	<30	22.4
[130]	0.5 μm CMOS	−28 dBm @ 1 Gb s^{-1}	<30	18
[130]	0.5 μm CMOS	−22.8 dBm @ 1.3 Gb s^{-1}	<30	8.0
[131]	0.6 μm CMOS	−20 dBm @ 1.25 Gb s^{-1}	85	
[86]	0.5 μm CMOS	−35 dBm @ 156 Mb s^{-1}	325	26.0
[8]	0.25 μm CMOS	−39.3 dBm @ 156 Mb s^{-1}	71.4	33.3
[132]	0.18 μm CMOS	−22 dBm @ 1.25 Gb s^{-1}		18.5
[3]	0.35 μm CMOS	−27 dBm @ 1.25 Gb s^{-1} [d]		21
[5]	0.18 μm CMOS	−29 dBm @ 1.25 Gb s^{-1} [d]		21
[2]	0.35 μm CMOS	−22.8 dBm @ 1.25 Gb s^{-1} [b]		
[2]	0.35 μm CMOS	−19.8 dBm @ 2.5 Gb s^{-1} [b]		
[134]	0.25 μm CMOS	−25 dBm @ 2.5 Gb s^{-1} [b]		
[136]	0.18 μm CMOS	−21.1 dBm @ 5 Gb s^{-1} [b]	47	
[137]	0.13 μm CMOS	−21 dBm @ 2 Gb s^{-1} [e]		
[137]	0.13 μm CMOS	−15 dBm @ 3.125 Gb s^{-1} [e]		
[137]	0.13 μm CMOS	−10.9 dBm @ 5 Gb s^{-1} [e]		
[137]	0.13 μm CMOS	+2 dBm @ 8 Gb s^{-1} [e]		
Sect. 9.3.1	0.18 μm CMOS	−30.7 dBm @ 622 Mb s^{-1}	97.2	
Sect. 9.3.1	0.18 μm CMOS	−23.7 dBm @ 1.25 Gb s^{-1}	97.2	
Sect. 9.3.3	120 nm CMOS	−31.4 dBm @ 622 Mb s^{-1}	87	22.8
Sect. 9.3.3	120 nm CMOS	−27.3 dBm @ 1.25 Gb s^{-1}	87	22.8
Sect. 9.3.2	120 nm CMOS	−29.9 dBm @ 622 Mb s^{-1}	154.5	19.3
Sect. 9.3.2	120 nm CMOS	−28.2 dBm @ 1.25 Gb s^{-1}	154.5	19.3
Sect. 9.3.4	120 nm CMOS	−31.3 dBm @ 622 Mb s^{-1}	88.5	29.7
Sect. 9.3.4	120 nm CMOS	−28.6 dBm @ 1.25 Gb s^{-1}	88.5	27
Sect. 9.3.4	120 nm CMOS	−14.5 dBm @ 2.5 Gb s^{-1}	88.5	12.9
Sect. 9.3.5	120 nm CMOS	−27.6 dBm @ 1.25 Gb s^{-1}	293	21
Sect. 9.3.5	120 nm CMOS	−20.4 dBm @ 2.5 Gb s^{-1}	293	13.8
Sect. 9.3.6	120 nm CMOS	−26 dBm [d]	227.5 [d]	19 [d]

[a] APD, $m = 10$

[b] Calculated value out of given input noise value and data of the photodiode described in the beginning of the chapter

[c] APD, $m = 6$

[d] Simulated result

[e] Integrated photodiode

at 2.5 Gb s^{-1} will be better when they are measured again with a new laser source, which will be built together with a burst-mode laser source, due to the fact that the incoming optical signal with the used laser was from bad quality at 2.5 Gb s^{-1} and therefore with an accurate input signal the sensitivity should be better. The simulated sensitivity result of the last-presented design probably will not be met in measurements, because all former chips did not

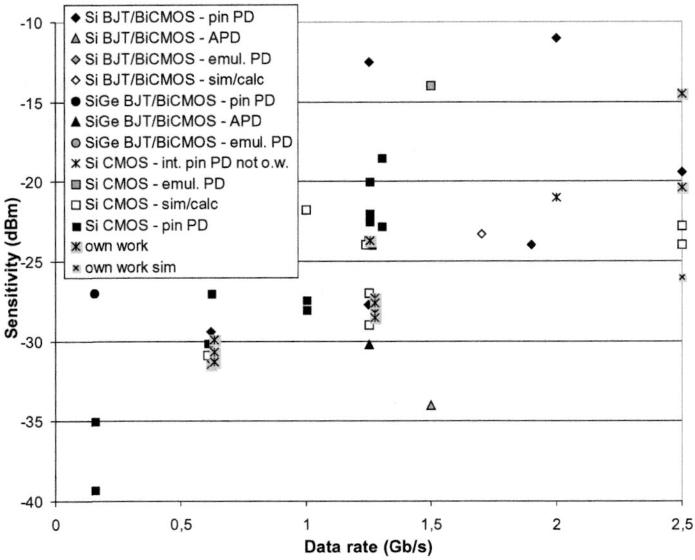

Fig. 9.52. Comparison of our own results with results known from literature

Table 9.9. Comparison of switching times between maximum and minimum optical input power

reference	switching time (ns)	maximum data rate
[86]	100	156 Mb s^{-1}
[8]	6.4	156 Mb s^{-1}
[132]	25	1.25 Gb s^{-1}
[3]	74	1.25 Gb s^{-1}
[4]	75	1.25 Gb s^{-1}
[5]	15–40s	1.25 Gb s^{-1}
[6]	25.6	1.25 Gb s^{-1}
three-stage BM TIA with internal feedback Sect. 9.3.4	13.4	2.5 Gb s^{-1}
three-stage BM TIA for 2.5 Gb s^{-1} Sect. 9.3.5	2.3	2.5 Gb s^{-1}
BM TIA for ultra-fast switching Sect. 9.3.6	1.64	2.5 Gb s^{-1}

meet the expectations from simulation concerning the sensitivity. Therefore a result in the range of −24 to −25 dBm is expected at 2.5 $\mathrm{Gb\,s^{-1}}$ instead of the simulated value of −26 dBm.

All presented TIA chips, except the folded-cascode TIA of Sect. 9.3.1, have a switchable transimpedance gain for a wide optical input power range of up to 27 dB. The variation of the feedback capacitance is avoided by a new approach of reducing the open-loop gain of the amplifier instead. Fast switching was achieved for the designs of the three-stage BM TIA with internal feedback (Sect. 9.3.4), of the three-stage BM TIA for 2.5 $\mathrm{Gb\,s^{-1}}$ described in Sect. 9.3.5, and of the BM TIA for ultra-fast gain switching, see Sect. 9.3.6. Table 9.9 shows a summary of the switching times of the presented results. The comparison with the results known from the literature shows that the only work within the same order of magnitude in switching time achieves with 156 $\mathrm{Mb\,s^{-1}}$ a considerably lower maximum data rate than our own work [8]. All switching times of the other designs are at least ten times slower than the best result of our own work.

References

1. R. Tao, M. Berroth, Wideband fully differential CMOS transimpedance preamplifier, IEE Electron. Lett. **21**, 1488 (2003)
2. W. Chen, C. Lu, A 2.5 Gb/s CMOS optical receiver analog front-end, in IEEE Custom Integrated Circuits Conference (2002), pp. 359–362
3. H.-S. Ko, Sub-Han, M.-S. Lee, S.-H. Chgai, 1.25Gb/s burst-mode optical receiver for the Ethernet PON using 0.35 μm CMOS technology, in 6th International Conference on Advanced Communications Technology, vol. 2 (2004), pp. 877–880
4. S. Han, M.-S. Lee, Burst-mode penalty of AC-coupled optical receivers optimized for 8B/10B line code, IEEE Photon. Technol. Lett. **16**(7), 1724 (2004)
5. J.-W. Seo, S. Han, S.-G. Lee, M.-S. Lee, T.W. Yoo, A 1.25Gb/s high sensitive peak detector in optical burst-mode receiver using a 0.18 μm CMOS technology, in *Proceedings of International Conference on Communication Technology (ICCT)*, 2003, pp. 644–646
6. P. Ossieur, Y.C. Yi, J. Bauwelinck, X.Z. Qiu, J. Verndewege, E. Gilon, DC-coupled 1.25 Gbit/s burst-mode receiver with automatic offset compensation, IEE Electron. Lett. **40**(7), 447 (2004)
7. E. Säckinger, Broadband circuits for optical fiber communication, in *El. Lab. Advanced Engineering Course "IC Design for Optical Communication Systems"* (EPFL, Lausanne, Switzerland, 2001)
8. S. Yamashita, S. Ide, K. Mori, A. Hayakawa, N. Ueno, K. Tanaka, Novel cell-AGC technique for burst-mode CMOS preamplifier with wide dynamic range and high sensitivity for ATM-PON system, in *Proceedings of 27th European Solid-State Circuits Conference (ESSCIRC)*, 2001, pp. 200–203
9. S.R. Nagel, Optical fibre—the expanding medium, IEEE Circuits Dev. Mag. **5**(2), 36 (1989)
10. W. Leeb, *Optische Nachrichtentechnik* (Institute of Communications and Radio-Frequency Engineering, TU-Vienna, Vienna, 1999)
11. E.D. Palik, *Handbook of Optical Constants of Solids* (Academic, Orlando, 1985), pp. 547–569
12. A. Ghazi, *Entwurf und Realisierung von integrierten Photodetektoren und CMOS-Schaltkreisen für optische Speichersysteme* (Wissenschaftsverlag Mainz, Aachen, 2000)

13. H. Zimmermann, *Silicon Optoelectronic Integrated Circuits* (Springer, Berlin Heidelberg New York, 2003)
14. H. Zimmermann, *Integrated Silicon Optoelectronics* (Springer, Berlin Heidelberg New York, 2000)
15. E. Kasper, H.J. Herzog, H. Kibbel, A one-dimensional SiGe superlattice grown by UHV epitaxy, Appl. Phys. **8**, 199 (1975)
16. H.M. Manasevit, I.S. Gergis, A.B. Jones, The properties of Si/$Si_{1-x}Ge_x$ films grown on Si substrates by chemical vapor deposition, J. Electron. Mater. **12**(4), 637 (1983)
17. J.C. Bean, T.T. Sheng, L.C. Feldman, A.T. Fiory, R.T. Lynch, Pseudomorphic growth of Ge_xSi_{1-x} on silicon by molecular beam epitaxy, Appl. Phys. Lett. **44**(1), 102 (1984)
18. R. People, J.C. Bean, D.V. Lang, A.M. Sergent, H.L. Stormer, K.W. Wecht, R.T. Lynch, K. Baldwin, Modulation doping in Ge_xSi_{1-x}/Si strained layer heterostructures, Appl. Phys. Lett. **45**(11), 1231 (1984)
19. R. People, Indirect band gap of coherently strained Ge_xSi_{1-x} bulk alloys on $\langle 001 \rangle$ silicon substrates, Phys. Rev. B **32**(2), 1405 (1985).
20. D.C. Ahlgren, M. Gilbert, D. Greenberg, S.J. Jeng, J. Malinowski, D. Nguyen-Ngoc, K. Schonenberg, K. Stein, R. Groves, K. Walter, G. Hueckel, D. Colavito, G. Freeman, D. Sunderland, D.L. Harame, B. Meyerson, Manufacturability demonstration of an integrated SiGe HBT technology for the analog and wireless marketplace, in *IEEE International Electronic Device Meeting, Washington, DC*, 1996, pp. 859–862
21. S. Subbanna, D. Ahlgren, D. Harame, B. Meyerson, How SiGe evolved into a manufacturable semiconductor production process, in IEEE International Solid-State Circuits Conference (1999), pp. 66–67
22. D.J. Paul, Silicon-germanium strained layer materials in microelectronics, Adv. Mater. **11**(3), 191 (1999)
23. S. Luryi, T.P. Pearsall, H. Tempkin, J.C. Bean, Waveguide infrared photodetectors on a silicon chip, IEEE Electron. Dev. Lett. **7**(2), 104 (1986)
24. A. Splett, T. Zinke, K. Petermann, E. Kasper, H. Kibbel, H.-J. Herzog, H. Presting, Integration of waveguides and photodetectors in SiGe for 1.3 μm operation, IEEE Photon. Technol. Lett. **6**(1), 59 (1994)
25. S.B. Samavedam, M.T. Currie, T.A. Langdo, E.A. Fitzgerald, High-quality germanium photodiodes integrated on silicon substrates using optimized relaxed graded buffers, Appl. Phys. Lett. **73**(15), 2125 (1998)
26. T.L. Lin, J. Maserjian, Characterization of wafer bonded photodetectors fabricated using various annealing temperatures and ambients, Appl. Phys. Lett. **57**(14), 1422 (1990)
27. T.L. Lin, E.W. Jones, A. Ksendzov, S.M. Dejewski, R.W. Fathauer, T.N. Krabach, J. Maserjian, A novel Si-based LWIR detector: the SiGe/Si heterojunction internal photoemission detector, in *IEDM Digest Technical Papers*, 1990, pp. 641–644
28. R.P.G. Karunasiri, J.S. Park, K.L. Wang, $Si_{1-x}Ge_x$/Si multiple quantum well infrared detector, Appl. Phys. Lett. **59**(20), 2588 (1991)
29. R.A. Soref, Silicon-based photonic devices, in *ISSCC Digest Technical Papers*, 1995, pp. 66–67
30. Q. Mi, X. Xiao, J.C. Sturm, L. Lenchyshyn, M.L.W. Thewalt, Room-temperature 1.3 μm electroluminescence from strained $Si_{1-x}Ge_x$/Si quantum wells, Appl. Phys. Lett. **60**(25), 3177 (1992)

31. H. Presting, T. Zinke, A. Splett, H. Kibbel, M. Jaros, Room-temperature electroluminescence from $Si/Ge/Si_{1-x}Ge_x$ quantum-well diodes grown by molecular-beam epitaxy, Appl. Phys. Lett. **69**, 2376 (1996)
32. R.A. Soref, Silicon-based optoelectronics, Proc. IEEE **81**(12), 1687 (1993)
33. G. Abstreiter, Engineering the future of electronics, Phys. World **5**, 36 (1992)
34. J.C. Sturm, Advanced column-IV epitaxial materials for silicon-based optoelectronics, MRS Bull. **April**, 60 (1998)
35. D.C. Houghton, Strain relaxation kinetics in $Si_{1-x}Ge_x/Si$ heterostructures, J. Appl. Phys. **70**(4), 2136 (1991)
36. D.C. Houghton, C. Gibbings, C. Tuppen, M. Lyons, M. Halliwell, Equilibrium critical thickness for $Si_{1-x}Ge_x$ strained layers on (100) Si, Appl. Phys. Lett. **56**(5), 460 (1990)
37. F.K. LeGoues, B.S. Meyerson, J. Morar, Anomalous strain relaxation in SiGe thin films and superlattices, Phys. Rev. Lett. **66**(22), 2903 (1991)
38. Y.J. Mii, Y.H. Xie, E.A. Fitzgerald, D. Monroe, F.A. Thiel, B.E. Weir, L.C. Feldman, Extremely high electron mobility in Si/Ge_xSi_{1-x} structures grown by molecular beam epitaxy, Appl. Phys. Lett. **59**(13), 1611 (1991)
39. K. Ismail, S. Rishton, J.O. Chu, K. Chan, B.S. Meyerson, High-performance Si/SiGe n-type modulation-doped transistors, IEEE Electron. Device Lett. **14**(7), 348 (1993)
40. G. Kissinger, T. Morgenstern, G. Morgenstern, H. Richter, Stepwise equilibrated graded Ge_xSi_{1-x} buffer with very low threading dislocation density on Si(001), Appl. Phys. Lett. **66**(16), 2083 (1995)
41. E.J. Prinz, P.M. Garone, P.V. Schwartz, X. Xiao, J.C. Sturm, The effects of base dopant outdiffusion and undoped $Si_{1-x}Ge_x$ junction spacer layers in $Si/Si_{1-x}Ge_x/Si$ heterojunction bipolar transistors, IEEE Electron. Dev. Lett. **12**(2), 42 (1991)
42. J.W. Slotboom, B. Streutker, A. Pruijimboom, D.J. Gravesteijn, Parasitic energy barriers in SiGe HBTs, IEEE Electron. Dev. Lett. **12**(9), 486 (1991)
43. U. Gnutzmann, K. Clausecker, Theory of direct optical transitions in an optical indirect semiconductor with a superlattice structure, Appl. Phys. **3**, 9 (1974)
44. E.D. Palik, *Handbook of Optical Constants of Solids II* (Academic, Boston, 1991), pp. 607–636
45. J.C. Bean, Recent developments in the strained layer epitaxy of germanium-silicon alloys, J. Vac. Sci. Technol. **B4**(6), 1427 (1986)
46. B. Jalali, A.F.J. Levi, F. Ross, E.A. Fitzgerald, SiGe waveguide photodetectors grown by rapid thermal chemical vapour deposition, Electron. Lett. **28**, 269 (1992)
47. B. Schuppert, J. Schmidtchen, A. Splett, U. Fischer, T. Zinke, R. Moosburger, K. Petermann, Integrated optics in silicon and SiGe-heterostructures, IEEE J. Lightwave Technol. **14**(10), 2311 (1996)
48. A. Vonsovici, Room temperature photocurrent spectroscopy of SiGe/Si p-i-n photodiodes grown by selective epitaxy, IEEE Trans. Electron. Dev. **45**(2), 538 (1998)
49. E.A. Fitzgerald, Y.H. Xie, M.L. Green, D. Brasen, A.R. Kortan, J. Michel, Y.J. Mii, B.E. Weir, Totally relaxed Ge_xSi_{1-x} layers with low threading dislocation densities grown on Si substrates, Appl. Phys. Lett. **59**(7), 811 (1991)
50. M.T. Currie, S.B. Samavedam, T.A. Langdo, C.W. Leitz, E.A. Fitzgerald, Controlling threading dislocation densities in Ge on Si using graded SiGe layers and chemical–mechanical polishing, Appl. Phys. Lett. **72**(14), 1718 (1998)

51. S. Luryi, A. Katalsky, J.C. Bean, New infrared detector on a silicon chip, IEEE Trans. Electron. Dev. **31**(9), 1135 (1984)
52. J. Oh, J.C. Campbell, S.G. Thomas, R.T.S. Bharatan, C. Jasper, R.E. Jones, T.E. Zirkle, Interdigitated Ge p-i-n photodetectors fabricated on a Si substrate using graded SiGe buffer layers, IEEE J. Quantum Electron. **38**(9), 1238 (2002)
53. M. Jutzi, M. Berroth, G. Whol, M. Oehme, V. Stefani, E. Kasper, Ge-on-Si pin-photodiodes for vertical and in-plane detection of 1300 to 1580 nm light, in *Proceeding of the 34th European Solid-State Device Research Conference (ESSDERC)*, 2004, pp. 345–348
54. Z. Huang, J. Oh, J.C. Campbell, Back-side-illuminated high-speed Ge photodetector fabricated on Si substrate using thin SiGe buffer layers, Appl. Phys. Lett. **85**(15), 3286 (2004)
55. J. Oh, S.K. Banerjee, J.C. Campbell, Metal–germanium–metal photodetectors on heteroepitaxial Ge-on-Si with amorphous Ge Schottky barrier enhancement layer, IEEE Photon. Technol. Lett. **16**(2), 581 (2004)
56. L. Colace, G. Masini, F. Galluzzi, G. Assanto, Metal–semiconductor–metal near-infrared light detector based on epitaxial Ge/Si, Appl. Phys. Lett. **72**(6), 3175 (1998)
57. L. Colace, G. Masini, G. Assanto, Ge-on-Si approaches to the detection of near-infrared light, IEEE J. Quantum Electron. **35**(12), 1843 (1999)
58. G. Masini, V. Cencelli, L. Colace, F. de Notaristefani, G. Assanto, Linear array of Si–Ge heterojunction photodetectors monolithically integrated with silicon CMOS readout electronics, IEEE J. Sel. Top. Quantum Electron. **10**(4), 811 (2004)
59. G. Dehlinger, S.J. Koester, J.D. Schaub, J.O. Chu, Q.C. Ouyang, A. Grill, High-speed Germanium-on-SOI lateral PIN photodiodes, IEEE Photon. Technol. Lett. **16**(11), 2547 (2004)
60. J. Oh, S. Csutak, J.C. Campbell, High-speed interdigitated Ge PIN photodetectors, IEEE Photon. Technol. Lett. **14**(3), 369 (2002)
61. S.J. Koester, J.D. Schaub, G. Dehlinger, J.O. Chu, Q.C. Ouyang, A. Grill, High-efficieny Ge-on-SOI lateral PIN photodiodes with 29 GHz bandwidth, in 62th Device Research Conference (DRC), vol. 1 (2004), pp. 175–176
62. O.I. Dosunmu, D.D. Cannon, M.K. Emsley, L.C. Kimerling, M.S. Ünlü, High-speed resonant cavity enhanced Ge photodetectors on reflecting Si substrates for 1550-nm operation, IEEE Photon. Technol. Lett. **17**(1), 175 (2005)
63. F.Y. Huang, X. Zhu, M.O. Tanner, Normal-incidence strained-layer superlattice $Ge_{0.5}Si_{0.5}$/Si photodiodes near 1.3 μm, Appl. Phys. Lett. **67**(4), 566 (1995)
64. H. Temkin, T.P. Pearsall, J.C. Bean, R.A. Logan, S. Luryi, Ge_xSi_{1-x} strained-layer superlattice waveguide photodetectors operating near 1.3 μm, Appl. Phys. Lett. **49**(13), 809 (1986)
65. H. Temkin, A. Antreasyan, N.A. Olsson, T.P. Pearsall, J.C. Bean, $Ge_{0.6}Si_{0.4}$ rib waveguide avalanche photodetectors for 1.3 μm operation, Appl. Phys. Lett. **49**(13), 809 (1986)
66. J.B. Posthill, R.A. Rudder, S.V. Hattangaddy, C.G. Fountain, R.J. Markunas, On the feasibility of growing dilute C_xSi_{1-x} epitaxial alloys, Appl. Phys. Lett. **56**(8), 734 (1990)
67. K. Eberl, S.S. Iyer, S. Zollner, J.C. Tsang, F.K. LeGoues, Growth and strain compensation effects in the ternary $Si_{1-x-y}Ge_xC_y$ alloy system, Appl. Phys. Lett. **60**(24), 3033 (1992)

68. Z.Y. Chang, *Low Noise Wide Band Amplifiers in Bipolar and CMOS Technologies* (Kluwer, Dordrecht, 1991)
69. Landolt-Börnstein, *Numerical Data and Functional Relationships in Science and Technology*, vol. 22a (Springer, Berlin Heidelberg New York, 1989)
70. R.A. Soref, Optical band gap of the ternary semiconductor $Si_{1-x-y}Ge_xC_y$, J. Appl. Phys. **70**(4), 2470 (1991)
71. L.C. Chang, A.S. Amour, J.C. Sturm, The effect of carbon on the valence band offset of compressively strained $Si_{1-x-y}Ge_xC_y$/(100)Si heterojunctions, Appl. Phys. Lett. **70**(12), 1557 (1997)
72. P.A. Stolk, D.J. Eaglesham, H.J. Gossmann, J.M. Poate, Carbon incorporation in silicon for suppressing interstitial-enhanced boron diffusion, Appl. Phys. Lett. **66**(11), 1370 (1995)
73. L.D. Lanzerotti, J.C. Sturm, E. Stach, R. Hull, T. Buyuklimanli, C. Magee, Suppression of boron transient enhanced diffusion in SiGe heterojunction bipolar transistors by carbon incorporation, Appl. Phys. Lett. **70**(23), 3125 (1997)
74. T. Tashiro, T. Tatsumi, M. Sugiyama, T. Hashimoto, T. Morikawa, A selective epitaxial SiGe/Si planar photodetector for Si-based OEIC's, IEEE Trans. Electron. Dev. **44**(4), 545 (1997)
75. R.T. Carline, D.A.O. Hope, V. Nayar, D.J. Robbins, M.B. Stanaway, A vertical cavity longwave infrared SiGe/Si photodetector using a buried silicide mirror, in *IEEE International Electron Device Meeting,* Washington, DC, 1997, pp. 891–894
76. A. Schüppen, U. Erben, A. Gruhle, H. Kibbel, H. Schumacher, U. König, Enhanced SiGe heterojunction bipolar transistors with 160 GHz-f_{max}, in *IEDM Digest Technical Papers*, 1995, pp. 743–746
77. J.-S. Rieh, D. Klotzkin, O. Qasaimeh, L.-H. Lu, K. Yang, L.P.B. Katehi, P. Bhattacharya, E.T. Croke, Monolithically integrated SiGe-Si PIN-HBT front-end photoreceiver, IEEE Photon. Technol. Lett. **10**(3), 415 (1998)
78. T. Yoon, B. Jalali, 622 Mb/s CMOS limiting amplifier with 40 dB dynamic range, IEE Electron. Lett. **32**(20) (1996)
79. E.M. Cherry, D.E. Hooper, The design of wide-band transistor feedback amplifiers, Proc. IEEE **110**, 375 (1963)
80. C. Holdenried, J. Haslett, M. Lynch, Analysis and design of HBT Cherry–Hooper amplifiers with emitter-follower feedback for optical communications, IEEE J. Solid-State Circuits **39**, 1959 (2004)
81. A. Maxim, A 10Gb/s SiGe transimpedance amplifier using a pseudo-differential input stage and a modified Cherry–Hooper amplifier, in *Digest of Technical Papers Symposium on VLSI Circuits*, 2004, pp. 404–407
82. T. Yoon, B. Jalali, Front-end CMOS chipset for fiber-based gigabit Ethernet, in *Digest of Technical Papers Symposium on VLSI Circuits*, 1998, pp. 188–191
83. Y. Ota, R.G. Swartz, Burst-mode compatible optical receiver with a large dynamic range, J. Lightwave Technol. **8**(12), 1897 (1990)
84. D. Yamazaki, N. Nagase, H. Nobuhara, T. Funaki, K. Wakao, 156 Mbit/s preamplifier IC with wide dynamic range for ATM-PON application, Electron. Lett. **33**(15), 1308 (1997)
85. M. Nakamura, N. Ishihara, Y. Akazawa, H. Kimura, An instantaneous response CMOS optical receiver IC with wide dynamic range and extremely high sensitivity using feed-forward auto-bias adjustment, IEEE J. Solid-State Circuits **30**(9), 991 (1995)

86. M. Nakamura, N. Ishihara, Y. Akazawa, A 156-Mb/s CMOS optical receiver for burst-mode transmission, IEEE J. Solid-State Circuits **33**(8), 1179 (1998)
87. H. Zimmermann, *Integrated Silicon Optoelectronics* (Springer, Berlin Heidelberg New York, 2000), pp. 29–40
88. J.D. Cressler, SiGe HBT technology: a new contender for Si-based RF and microwave circuit applications, IEEE Trans. Microwave Theory Tech. **46**, 572 (1998)
89. J. Dunn et al., Foundation of rf CMOS and SiGe BiCMOS technologies, IBM J. Res. Dev. **47**(2/3), 101 (2003)
90. D. Greenberg, S. Sweeney, C. LaMothe, K. Jenkins, D. Friedman, B. Martin Jr., G. Freeman, D. Ahlgren, S. Subbanna, A. Joseph, Noise performance and considerations for integrated RF/analog/mixed-signal design in a high-performance SiGe BiCMOS technology, in *International Electron Devices Meeting (IEDM) Technical Digest Papers, Digital Object Identifier 10.1109/IEDM.2001.979553*, 2001, pp. 22.1.1–22.1.4
91. D.L. Harame, D.C. Ahlgren, D.D. Coolbaugh, J.S. Dunn, G.G. Freeman, J.D. Gillis, R.A. Groves, G.N. Hendersen, R.A. Johnson, A.J. Joseph, S. Subbanna, A.M. Victor, K.M. Watson, C.S. Wbster, P.J. Zampardi, Current status and future trends of SiGe BiCMOS technology, IEEE Trans. Electron. Dev. **48**(11), 2575 (2001)
92. International Technology Roadmap for Semiconductors. http://public.itrs.net/, 2003
93. Infineon Technologies Robust Digital Circuits and Signal Processing. http://www.infineon.com/cmc_upload/robust_digital_circuits.pdf, 2004
94. J. Glaser, *Burst Mode Receiver for Passive Optical Networks*, Diploma Thesis, TU-Vienna, 2004
95. R.C. Jaeger, *Microelectronic Circuit Design*, (McGraw-Hill, New York, 1997), p. 993
96. S.M. Park, C. Papavassiliou, On the design of low-noise, Giga-Hertz bandwidth preamplifiers for optical receiver applications, in *Proceedings of ICECS*, vol. 2 (1999), pp. 785–788
97. N. Haralabidis, D. Loukas, K. Misiakos, S. Katsafouros, A transimpedance CMOS multichannel amplifer with a $50\,\Omega$ wide-output-range buffer for high counting rate applications, IEEE J. Solid-State Circuits **32**, 135 (1994)
98. K. Schrödinger, J. Stimma, M. Mauthe, A fully integrated CMOS receiver front-end for optic Gigabit Ethernet, IEEE J. Solid-State Circuits **7**, 874 (2002)
99. K. Schneider, H. Zimmermann, Highly sensitive wide-dynamic range optical receivers, in *Proceedings of* 12*th International Conference on Mixed Design of Integrated Circuits and Systems (MIXDES)*, vol. 2, 2005, pp. 953–958
100. A. Weinmann, *Regelungen – Analyse und technischer Entwurf Band 1: Systemtechnik linearer und linearisierter Regelungen auf anwendungsbasierter Grundlage, 3. Auflage* (Springer, Berlin Heidelberg New York, 1994)
101. K.J. Ebeling, *Integrated Optoelectronics* (Springer, Berlin Heidelberg New York, 1993)
102. K. Schneider, H. Zimmermann, Folded-cascode transimpedance amplifier for burst-mode applications, in *Proceedings of 10th IEEE International Symposium on Electron Devices for Microwave and Optoelectronic Applications (EDMO)*, 2002, pp. 294–299
103. P.R. Gray, P.J. Hurst, S.H. Lewis, R.G. Meyer, *Analysis and Design of Analog Integrated Circuits* (Wiley, New York, 2001), pp. 748–807

104. P.R. Gray, P.J. Hurst, S.H. Lewis, R.G. Meyer, *Analysis and Design of Analog Integrated Circuits* (Wiley, New York, 2001), pp. 78–169
105. T. Takeshita, T. Nishimura, 622 Mb/s fully-integrated optical IC with a wide range input, IEEE International Solid-State Circuits Conference (2002), pp. 258–259
106. Philips Semiconductors, *Data Sheet of TZA3043, Gigabit Ethernet/Fibre Channel transipmedance amplifier.* http://www.semiconductors.philips.com, 2002
107. Y. Ota, R.G. Swartz, V.D.I. Archer, S.K. Korotky, M. Banu, A.E. Dunlop, High-speed, burst-mode, packet-capable optical receiver and instantaneous clock recovery for optical bus operation, J. Lightwave Technol. **12**(2), 325 (1994)
108. G. Halkias, N. Haralabidis, E.D. Kyriakis-Bitzaros, S. Katsafouros, 1.7 GHz bipolar otpoelectronic receiver using conventional 0.8 μm BiCMOS process, in *Proceedings of International Symposium on Circuits and Systems (ISCAS)*, vol. 5, 2000, pp. 417–420
109. V. Ramakrishnan, J.N. Albers, R.N. Nottenburg, W.J. Hillery, Monolithic re-timed receiver array with dynamic reference level compensator for 4×2 Gb/s synchronous optical data link, IEEE Photon. Technol. Lett. **8**(1), 82 (1997)
110. M. Hirose, N. Ishihara, Y. Akazawa, H. Ichino, An ultracompact, 2-cc-size, low-power 2.5 Gb/s optical receiver module incorporating an MU receptacle, J. Lightwave Technol. **17**(11), 2349 (1999)
111. J. Hauenschild, D. Friedrich, J. Herrle, J. Krug, PIN-preamp module and CDR-DEMUX constituting a receiver for short-haul links up to 3.5 Gb/s, IEEE J. Solid-State Circuits **33**(12), 2247 (1998)
112. T. Nagahori, K. Miyoshi, Y. Aizawa, Y. Kusachi, Y. Nukada, N. Kami, N. Suzuki, An analog front-end chip set employing an electro-optical mixed design on SPICE for 5 Gb/s/ch parallel optical interconnection, IEEE J. Solid-State Circuits **36**(12), 1984 (2001)
113. H.H. Kim, S. Chandrasekhar, C. Burrus, J. Baumann, A Si BiCMOS transimpedance amplifier for 10-Gb/s SONET receiver, IEEE J. Solid-State Circuits **36**(5), 769 (2001)
114. H.H. Kim, S. Chandrasekhar, C. Burrus, J. Baumann, *An Si BiCMOS Transimpedance Amplifier for 10 Gb SONET Receiver* (2000), pp. 170–173
115. R. Takeyari, K. Ishikawa, T. Masuda, S. Ueno, T. Harada, K. Harada, K. Ohhata, N. Endoh, H. Chiba, 10 Gb/s Si receiver for short/long haul transmission, in International Conference on Integrated Optics and Optical Fibre Communicatons and European Conference on Optical Communication, vol. 4 (1997)
116. Y. Lu, M.N. El-Gamal, A 2.3 V low noise, low power 10 GHz bandwidth Si-bipolar transimpedance preamplifier for optical receiver front-ends, in *IEEE International Symposium on Cirucuits and Systems (ISCAS)*, vol. 4, 2001, pp. 834–837
117. M. Nakamura, M. Okayasu, N. Ishihara, 5 V, 10 Gb/s optical receiver module using Si-bipolar IC, Electron. Lett. **36**(25), 2085 (2000)
118. M. Neuhäuser, H.-M. Rein, H. Wernz, A. Felder, 13 Gb/s Si bipolar preamplifier for optical front ends, Electron. Lett. **29**(5), 492 (1993)
119. S. Brigati, P. Colombara, L. D'Ascoli, U. Gatti, T. Kerekes, P. Malcovati, A. Profumo, A SiGe BiCMOS burst-mode 155 Mb/s receiver for PON, IEEE J. Solid-State Circuits **37**, 887 (2002)

120. S.M. Park, S. Hong, A 65 mW 5 Gb/s/ch current-mode common-base transimpedance amplifier array for optical interconnects, IEEE Photon. Technol. Lett. **15**(8), 1138 (2003)
121. S.M. Park, Four-channel SiGe transimpedance amplifier array for parallel optical interconnects, in *Proceedings of International Symposium on Circuits and Systems (ISCAS)*, vol. 4, 2004, pp. 213–216
122. A. Schild, H.-M. Rein, J. Mullrich, L. Altenhain, J. Blank, K. Schrodinger, High-gain SiGe transimpedance amplifier array for a 12×10 Gb/s parallel optical-fiber link, IEEE J. Solid-State Circuits **38**(1), 4 (2003)
123. B. Yang, J.D. Schaub, S.M. Csutak, D.L. Rogers, J.C. Campbell, 10-Gb/s all-silicon optical receiver, IEEE Photon. Technol. Lett. **15**(5), 745 (2003)
124. H. Matsuda, A. Miura, Y. Okamura, H. Irie, K. Ito, T. Toyonaka, H. Takahashi, T. Harada, High performance of 10-Gb/s APD/preamplifier optical-receiver module with compact size, IEEE Photon. Technol. Lett. **15**(2), 278 (2003)
125. M. Soda, H. Tezuka, F. Sato, T. Hashimoto, S. Nakamura, T. Tatsumi, T. Suzaki, T. Tashiro, Si-analog IC's for 20Gb/s optical receiver, IEEE J. Solid-State Circuits **29**(12), 1577 (1994)
126. T. Masuda, K. Ohhata, K. Oda, M. Tanabe, H. Shimamoto, T. Onai, K. Washio, 40Gb/s analog IC chipset for optical receiver using SiGe HBTs, in *ISSCC Digest Technical Papers*, 1998, pp. 314–315, 454
127. B. Razavi, A 622 Mb/s 4.5 pA/$\sqrt{Hz}$ CMOS transimpedance amplifier, in *IEEE International Solid-State Circuits Conference (ISSCC), Digest of Technical Papers* (2000), pp. 162–163
128. M. Ingels, M. Steyart, A 1-Gb/s, 0.7-μm CMOS optical receiver with full rail-to-rail output swing, IEEE J. Solid-State Circuits **34**(7), 971 (1999)
129. T.K. Woodward, A.V. Krishnamoorthy, R.G. Rozier, A.L. Lentine, Low-power, small-footprint gigabit Ethernet compatible optical receiver circuit in 0.25 μm CMOS, Electron. Lett. **36**(17), 1489 (2000)
130. T. Nakahara, H. Tsuda, N. Ishihara, K. Tateno, C. Amano, High-sensitivity 1 Gbit/s CMOS receiver integrated with GaAs- or InGaAs-photodiode by wafer-bonding, Electron. Lett. **37**(12), 781 (2001)
131. S.M. Park, H.-J. Yoo, 1.25 Gb/s regulated cascode CMOS transimpedance amplifier for Gigabit Ethernet applications, IEEE J. Solid-State Circuits **39**(1), 112 (2004)
132. Q. Le, S.-G. Lee, Y.-H. Oh, H.-Y. Kang, T.-H. Yoo, Burst-mode receiver for 1.25Gb/s ethernet PON with AGC and internally created reset signal, in *Proceedings of IEEE International Solid-State Circuits Conference* (2004), pp. 474–475
133. J. Lee, J. Park, J. Lim, K. Jung, S. Lim, Y. Cho, S.M. Park, J. Choi, A 1.25 Gb/s CMOS burst-mode optical receiver with automatic level restoration for PON applications, in *Midwest Symposium on Circuits and Systems (MWSCAS)*, vol. 1, 2004, pp. 137–140
134. P. Mitran, F. Beaudoin, M.N. El-Gamal, A 2.5 Gb/s CMOS optical receiver frontend, in *IEEE International Symposium on Cirucits and Systems (ISCAS)*, vol. 5, 2002, pp. 441–444
135. S.M.R. Hasan, A high performance wide-band CMOS transimpedance amplifier for optical transceivers, in *Proceedings of the 3rd IEEE International Worshop on System-on-Chip for Real Time Applications*, 2003, pp. 82–85

136. F. Beaudoin, M.N. El-Gamal, A 5-Gb/s CMOS optical receiver frontend, in *Midwest Symposium on Circuits and Systems (MWSCAS)*, vol. 3 (2002), pp. 168–171
137. S.M. Csutak, J.D. Schaub, W.E. Wu, J.C. Campbell, 8 Gb/s CMOS compatible monolithically integrated silicon optial receiver, in Optical Fiber Communication Conference and Exihibit (2002), pp. 585–586
138. W.-Z. Chen, Y.-L. Cheng, A 1.8V, 10 Gbps fully integrated CMOS optical receiver analog front end, in *ESSCIRC*, 2004, pp. 263–266
139. Device Research Group of the Department of of Electrical Engineering and Computer Science, University of California, Berkeley. http://www.device.eecs.berkeley.edu/~ bsim3/BSIM4/BSIM400/slide/slide.pdf, 2004
140. Device Research Group of the Department of of Electrical Engineering and Computer Science, University of California, Berkeley. http://www.device.eecs.berkeley.edu/~ ptm/130nm_nominal.txt, 2004
141. T. Yoon, B. Jalali, 1 Gbit/s fibre channel CMOS transimpedance amplifier, IEE Electron. Lett. **7**, 588 (1997)
142. H. Zimmermann, R. Swoboda, K. Schneider, J. Knorr, Comparison of CMOS and BiCMOS optical receiver SoCs, in *Conference on VLSI Circuits and Systems, Proceedings SPIE 5117*, 2003, pp. 598–609
143. K. Schneider, H. Zimmermann, A. Wiesbauer, Optical receiver in deep-submicrometer CMOS with −28.2 dBm sensitivity at 1.25 Gbit/s, IEE Electron. Lett. **40**(4), 262 (2004)
144. K. Schneider, H. Zimmermann, Low-noise 0.12 μm CMOS transimpedance amplifier, J. Circuits Syst. Comput. (JCSC) (accepted)
145. K. Phang, D. Johns, A CMOS optical preamplifier for wireless infrared communications, IEEE Trans. Circuits Syst. **46**(7), 852 (1999)
146. K. Schneider, H. Zimmermann, Low-noise transimpedance amplifier for optical receiver in 120 nm CMOS technology, in *Proceedings of the International Symposium on Signals, Systems, and Electronics (ISSSE)*, 2004, pp. 70–73
147. K. Schneider, H. Zimmermann, Three-stage transimpedance amplifier in deep-sub-μm CMOS technology, in *Asia Pacific Microwave Conference, Book of Abstracts* (2004), p. 326
148. K. Schneider, H. Zimmermann, Burst-mode receivers for passive opticla networks, in *Proceedings of the 12th IEEE Internhational Symposium on Electron Devices for Microwave and Optoelectronic Applications (EDMO)*, 2004, pp. 60–65
149. U. Tietze, C. Schenk, *Electronic Circuits: Design and Applications* (Springer, Berlin Heidelberg New York, 2002)

Index

absorption coefficient, 13
Affirma Analog Artist, 139
antireflection coating, 14
APD, 18, 115, 125
ATM-PON, 6
automatic gain control, 8
avalanche photodidoe, *see* APD

back-annotation, 139
back-gate effect, 47
bandgap energy, 43
bandwidth, 22, 35, 36, 85, 98, 100, 105, 108, 112, 113, 146, 151, 152, 156, 157, 160, 164, 165, 167, 169, 170
bandwidth-length product, 10
base current, 93
base resistor, 94
base-after-gate production, 45
base-during-gate production, 45
basim3v3, 141
basim4, 141
BER, 85–87, 146
bipolar transistor
  equivalent input noise current, 95
  equivalent input noise current density, 96
  equivalent input noise sources, 100
  equivalent input noise voltage, 94
  output noise current, 94
bipolar transistor noise model, 93
bit-AGC, 38
bit-error analyzer, 88
bit-error rate, *see* BER
bottom-level detector, 38

bsim3v3
  NMOS, 143
  PMOS, 144
bsim4, 142
burst-mode communication, 7

cell-AGC, 38
chemical mechanical polishing, 21
Cherry–Hooper architecture, 37
CMOS process variations, 49
CMP, 21
collector current, 94
collector series resistance, 93
communication analyzer, 87, 88
continuous-mode communication, 7
critical thickness, 20
cross section
  deep-sub-micron CMOS, 46
  SiGe heterojunction bipolar transistor, 43

decision threshold, 86
deep-sub-micron CMOS, 45, 46
dielectric function, 13
digital sampling oscilloscope, 87
DRC, 139

electronic noise, 85
equivalent input noise current spectral density, 100
ESD-structure, 147
extinction coefficient, 13
extinction ratio, 8, 90
  correction factor, 90

extrinsic base, 93

feedback network, 51
FET
  equivalent input noise current, 98
  equivalent input noise current density, 98
  equivalent input noise sources, 97, 101
  equivalent input noise voltage, 97
  small-signal equivalent circuit, 97
FET noise model, 96
fiber-to-the-home, 6
flicker noise, 92
folded-cascode amplifier
  equivalent input noise sources, 104
folded-cascode TIA, 148
  input noise current density, 107
  noise analysis, 104

Ge photodiodes, 22
graded buffer, 21
graded composition buffer, 21

HBT, 19, 32
heterojunction bipolar transistor, 19, 32
heterojunction bipolar transistor noise model, 93
high-V_t transistors, 47
histogram, 88

ideal TIA, 99
input resistance, 55
intermodal dispersion, 10
intrinsic base, 93
IRTS, 46

limiting amplifier, 8
low-V_t transistors, 47
low-absorption windows, 11
LVS, 139

main amplifier, 8
MOSFET model, 141

n-FET in n-well capacitors, 48
noise distance, 87
noise models, 91
  resistor noise model, 92
numerical aperture, 10

OEIC, 32
optical fibers, 9
  multimode fibers, 9, 10
    graded index, 10
    stepped index, 10
  single-mode fibers, 9
optical input power, 89, 90, 101
optical power meter, 146
optical receiver
  BiCMOS, 115, 116, 118
  burst-mode (BM), 115, 122, 129–131
  CMOS, 126–131, 133–135
  continuous-mode (CM), 115, 116, 119, 123, 125–129, 133–135
  Si bipolar, 115–120
  SiGe BiCMOS, 122, 124
  SiGe HBT, 121, 123, 125
optimized relaxed buffer, 21
ORB, 21
output resistance, 56

PCB, 145
photodiode, 8
photodiode model, 139
pin photodiode, 33, 116
pin-amplifier circuit, 101
plastic optical fibers, 11
pn photodiode, 21
point-to-multipoint network, 6
point-to-point access, 5
power penalty, 85, 89
PRBS, 146
printed circuit board, *see* PCB
probability density, 86
pseudorandom bit sequence, *see* PRBS

Q-factor, 87–89, 149
quantum efficiency, 13, 89
  dynamical quantum efficiency, 15, 17
  external quantum efficiency, 13
  internal quantum efficiency, 15
  optical quantum efficiency, 14
  stationary quantum efficiency, 15

reflectivity, 13
refractive index, 10
regular threshold, 47
regular-V_t transistors, 48
responsivity, 15, 22, 89
rf-CMOS transistors, 142

sensitivity, 85, 89, 101
shot noise, 91
shunt–shunt feedback, 51
SiGe HBT, 32, 43–45, 96, 102
SiGe HBT receiver, 121
silicon-on-isolator pin photodiode, 124
simulation environment, 139
  top-level, 141
single-mode fibers, 10
SoC, 146
spectre simulator, 139
switching time, 138
system-on-a-chip, *see* SoC

thermal noise, 92, 99
three-inverter TIA, 151
  input referred noise current density, 110
  noise analysis, 109
three-stage BM TIA for 2.5 $\mathrm{Gb\,s^{-1}}$, 163
three-stage BM TIA for ultra-Fast gain switching, 169
three-stage BM TIA with internal feedback, 159
three-stage TIA, 154
TIA noise model, 99
transmission bandwidth, 10
transresistance amplifier, 51, 56

upper bound for noise, 99

Virtuoso Layout Editor, 139
Virtuoso Schematic Editor, 139

y-parameters, 51

Springer Series in

ADVANCED MICROELECTRONICS

1 **Cellular Neural Networks**
Chaos, Complexity and VLSI Processing
By G. Manganaro, P. Arena, and L. Fortuna

2 **Technology of Integrated Circuits**
By D. Widmann, H. Mader, and H. Friedrich

3 **Ferroelectric Memories**
By J.F. Scott

4 **Microwave Resonators and Filters for Wireless Communication**
Theory, Design and Application
By M. Makimoto and S. Yamashita

5 **VLSI Memory Chip Design**
By K. Itoh

6 **Smart Power ICs**
Technologies and Applications
Ed. by B. Murari, R. Bertotti, and G.A. Vignola

7 **Noise in Semiconductor Devices**
Modeling and Simulation
By F. Bonani and G. Ghione

8 **Logic Synthesis for Asynchronous Controllers and Interfaces**
By J. Cortadella, M. Kishinevsky, A. Kondratyev, L. Lavagno, and A. Yakovlev

9 **Low Dielectric Constant Materials for IC Applications**
Editors: P.S. Ho, J. Leu, W.W. Lee

10 **Lock-in Thermography**
Basics and Use for Functional Diagnostics of Electronic Components
By O. Breitenstein and M. Langenkamp

11 **High-Frequency Bipolar Transistors**
Physics, Modelling, Applications
By M. Reisch

12 **Current Sense Amplifiers for Embedded SRAM in High-Performance System-on-a-Chip Designs**
By B. Wicht

13 **Silicon Optoelectronic Integrated Circuits**
By H. Zimmermann

14 **Integrated CMOS Circuits for Optical Communications**
By M. Ingels and M. Steyaert

15 **Gettering Defects in Semiconductors**
By V.A. Perevostchikov and V.D. Skoupov

16 **High Dielectric Constant Materials**
VLSI MOSFET Applications
Editors: H.R. Huff and D.C. Gilmer

17 **System-level Test and Validation of Hardware/Software Systems**
By M. Sonza Reorda, Z. Peng, and M. Violante

GPSR Compliance

The European Union's (EU) General Product Safety Regulation (GPSR) is a set of rules that requires consumer products to be safe and our obligations to ensure this.

If you have any concerns about our products, you can contact us on ProductSafety@springernature.com

In case Publisher is established outside the EU, the EU authorized representative is:

Springer Nature Customer Service Center GmbH
Europaplatz 3
69115 Heidelberg, Germany

Batch number: 10607182

Printed by Printforce, the Netherlands